COASTAL AND OCEANIC LANDFORMS, DEVELOPMENT AND MANAGEMENT

CORAL REEF ECOSYSTEM IN SPACE AND TIME

BASED ON THE REEFS OF VIETNAM

Coastal and Oceanic Landforms, Development and Management

Additional books in this series can be found on Nova's website under the Series tab.

Additional e-books in this series can be found on Nova's website under the eBooks tab.

COASTAL AND OCEANIC LANDFORMS, DEVELOPMENT AND MANAGEMENT

CORAL REEF ECOSYSTEM IN SPACE AND TIME

BASED ON THE REEFS OF VIETNAM

Russian Academy of Sciences
A. V. Zhirmunsky Institute of Marine Biology Far Eastern Branch

YURI LATYPOV

New York

Library of Congress Cataloging-in-Publication Data

Names: Latypov, kIIU. kIIA. (kIIUrifi kIIAkovlevich), editor.
Title: Coral reef ecosystem in space and time : (based on the reefs of Vietnam) / Yuri Latypov (A.V. Zhirmunsky Institute of Marine Biolology), editor.
Description: Hauppauge, New York : Nova Science Publishers, Inc., [2016] | Series: Coastal and oceanic landforms, development and management | Includes bibliographical references and index.
Identifiers: LCCN 2016003448 | ISBN 9781634847056 (softcover)
Subjects: LCSH: Coral reef ecology--Southeast Asia. | Coral reef ecology--Vietnam. | Coral reefs and islands--Southeast Asia. | Coral reefs and islands--Vietnam. | Aquatic ecology--Southeast Asia. | Aquatic ecology--Vietnam.
Classification: LCC QH193.S6 C67 2016 | DDC 577.7/890959--dc23 LC record available at http://lccn.loc.gov/2016003448

Published by Nova Science Publishers, Inc. † New York

CONTENTS

PREFACE

This lavishly illustrated book explores the concept of reef ecosystems and its characteristics. It provides a conceptual description of reefs and their functions. This compilation also outlines the general species composition and structure of coral reefs and their corallobionts. Described and illustrated are the main inhabitants of the reef community. Various types of reefs (fringing, barrier, platform etc.) in different regions of Vietnam, along with their conditions and statuses are catalogued as well. Reef ecosystems traced under anthropogenic influence and the impact of typhoons are included. This book shows the possibilities of artificially cultivating corals and the rebuilding of their communities.

Chapter 1

CONCEPTUAL INTRODUCTION

Fisheries and mariculture (plenty of fish, shrimp, lobster, fishing clams, pearls and seaweed) are undergoing constant development. The *populations* of the coastal waters of the Indo-Pacific Equatorial zone are linked to significant productivity, as determined by the study of coral reefs ecosystems..

Clear turquoise water ebbs and flows at the edge of a tropical shore. Beneath the water's surface exists a breathtaking underwater world. Home to a diversity of colorful exotic fish, corals, and countless other marine creatures - this is the coral reef. Oceans, seas and fresh water cover more than seventy percent of the earth's surface. While coral reefs take up only a very small fraction of the ocean (less than one-tenth of a percent), they are home to an astonishing variety of animals and plants. Coral reefs have very high biological diversity (biodiversity) - approximately 93,000 species of plants and animals have already been identified in coral reefs, and scientists predict that there may be over three million.

Coral reefs are the primary source of food and income for millions of people. They produce valuable chemical compounds for medicines, and provide natural wave barriers that protect beaches and coastlines from storms and floods. Yet coral reefs are in danger. Already, eleven percent of the world's coral reefs have been lost and another sixteen percent were severely damaged during the 1998 El Niño event. Scientists predict that another thirty-two percent may be lost in the next thirty years if human threats are not reduced. As our awareness of the value of coral reefs increases, so do our efforts to reduce current threats. Coral reef marine protected areas (MPAs), Integrated Coastal Zone Management (ICZM), sustainable tourism, education and outreach programs and coral reef rehabilitation are just a few of the many steps being taken to conserve and protect these valuable and beautiful ecosystems.

More and more species that live on coral reefs have been found to contain compounds that can be used in medicine (biomedical compounds), including some applied to the treatment of human immunodeficiency virus (HIV), cancer, ulcers and cardiovascular diseases. In addition, the unique skeletal structure of coral has been used to make our most advanced forms of bone grafting materials.

Over a thousand coral species exist worldwide. Stony (*hermatypic*) corals are the best recognized because of their elaborate and colorful formations. One trait of stony corals is their capacity to build reef structures that range from tens, to thousands of meters across. As they grow, reefs provide structural *habitats* for many different vertebrate and invertebrate species – a single reef may host tens of thousands of different species. Although corals are found throughout the world, reef-building corals are confined to waters that exhibit a narrow band of characteristics. The water must be warm, relatively clear, and saline. These waters are usually nutrient-poor as well. Physiologically and behaviorally, corals have evolved to take advantage of this unique environment and thrive.

Not only does a specific range of environmental conditions confine reef-building corals, but also as adults, almost all of them are *sessile*. This means that they remain on the same spot of the sea floor for their entire lives. Thus, reef-building corals have developed reproductive, feeding, and social behaviors that allow them to gain the maximum survival benefit from their situation.

Spreading colonies of asexually reproducing coral polyps provide shelter for a moray eel. Over the eons, many corals have evolved with the ability to reproduce both asexually and sexually. In asexual reproduction, new clonal polyps bud off from parent polyps to expand or begin new colonies [55]. This occurs when the parent polyp reaches a certain size and divides. The process continues throughout the animal's life, forming an ever-expanding colony [4]. It is worth noting that it is a judgment call whether to call this reproduction, since that implies that a polyp is an individual. In a coral, an individual could be either a polyp or the whole colony, and some lines of evidence indicate that the colony is the individual, not the polyp. Evidence for coral as the individual argument includes the polyps of a colony being connected by a nervous system and gastrovacular cavity, and becoming sexually reproductive when a colony reaches a certain size, not when a polyp reaches a certain size.

Corals can also reproduce asexually by fragmentation – that is, when a portion of the colony (say, a branch), is detached from the rest and falls into a suitable substrate. This can happen either naturally, perhaps when wave action from a storm breaks off a coral piece and settles it elsewhere, or when humans purposely take coral fragments and place them in other substrate areas [47, 17].

Already during the brief acquaintance with molecular bases of life, we encounter lipids. Let's call them the basic biological functions: the main components of bio membranes. spare, isolating and protecting material bodies; the most energetic part of food; an important part of the diet of humans and animals; carriers of a number of vitamins; water and salt transport regulators; immuno modulators; regulators of the activity of some enzymes; *endo-hormones*; transmitters of biological signals.

This list characterizes the study of lipids. In providing these and other features, lipids have different structures in different quantities: tons of triglycerides protect whales from external influences, and endo-hormones or transmitters of biological signals are lipids of other classes in the micro-and nanograms doses. Therefore, lipids must be pictured on the same level as proteins, nucleic acids and carbohydrates in order to understand many biological processes.

Chapter 2

CHARACTERISTIC OF CORALS

2.1. THE BIOLOGY OF CORALS

Because corals are sessile, they were for a long time thought to be plants. In Ovid's Metamorphoses he refers to coral as an organism that is soft under water but hardens on contact with air. (What he was actually seeing was the death of the living tissue, which exposed the hard skeleton.) In 1723, the naturalist Jean Andre Peyssonel proposed to the French Academy of Sciences that corals are animals. His view was derided, and he subsequently abandoned scientific work. Since then, of course, he has been proved right. Corals belong to the large and varied phylum of COELENTERATES, which are simple multicellular animals. The phylum's name is from the Greek kailos, hollow, and *enteron*, gut, because the main body cavity of its members is the digestive cavity.

The closest relatives of the true corals are the sea anemones, which corals resemble in basic body structure and overall appearance. The soft coral polyp consists of three layers of cells and a contractile sac crowned with a ring of six tentacles (or a multiple of six) surrounding a mouth-like opening. The tentacles have the specialized stinging cells called nematocysts, which discharge an arrow-like barb and a toxin that stuns animal prey such as microscopic crustaceans. From the mouth of the polyp the short muscular gullet descends into the stomach cavity and is connected to the body wall by six partitions (or a multiple of six), increasing the area of the digestive surface. The free edges of the partitions are extended into mesenterial filaments: convoluted tubes that can be extruded through the mouth or the body wall.

The size of the polyps is highly variable, from about one millimeter in diameter in some species to more than 20 centimeters in others. Each polyp can give rise to a large colony by asexual division, or budding. Corals also reproduce sexually, producing free-swimming larvae that settle and establish new colonies. The most striking feature of coral colonies is their ability to form a massive calcareous skeleton. Individual coral colonies weighing several hundred tons and large enough to fill a living room are common in many reefs. In most species, the polyps are in individual skeletal cups, some extending their tentacles to feed by night and some partially withdrawing into the cups by day. In the contracted condition, the polyps can resist drying or mechanical injury at low tide, when some of the colonies may be exposed. The skeletal cups consist of fan-shaped clusters of calcium carbonate crystals, which are arranged in patterns that are characteristic of each coral species (Figure 1-3).

A remarkable feature of all reef-building corals is their symbiosis with the unicellular algae known as *zooxanthelae*. The coral polyps contain large numbers of these algae within cells in the lining of their gut. The zooxanthelae are yellow-brown marine algae of the family DINOPHYCEAE, to which many of the free-living *dinoflagellate* algae also belong.

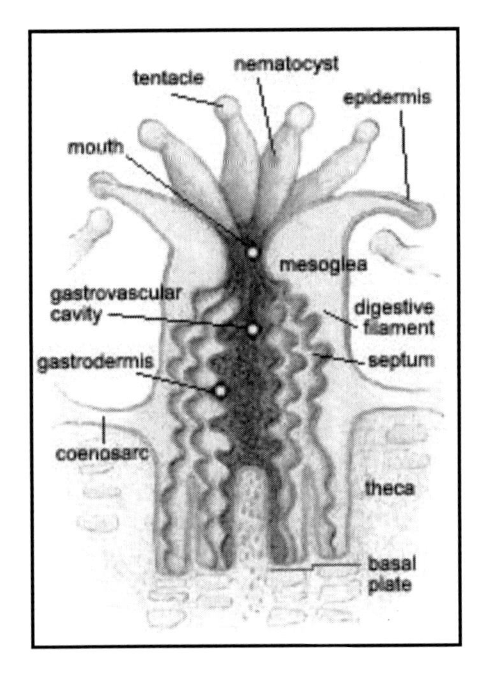

Figure 1. Schematic picture of coral polyp.

The algae live, conduct photosynthesis and divide within the cells of their coral host, and on this symbiosis rests the entire biological productivity of the coral-reef ecosystem, since the zooxanthelae of reef-building corals need light.

The corals also require warm waters (above 20 degrees Celsius) and do not tolerate low salinity or high turbidity. Where deeper colonies are shaded by a dense overgrowth of shallower ones, the deeper colonies maximize their light-gathering capacity by growing in ramifications like the branches of forest trees.

Figure 2. The appearance of the polyp with tentacles and slit-like mouth for photosynthesis, such corals grow only in ocean waters less than 100 meters deep.

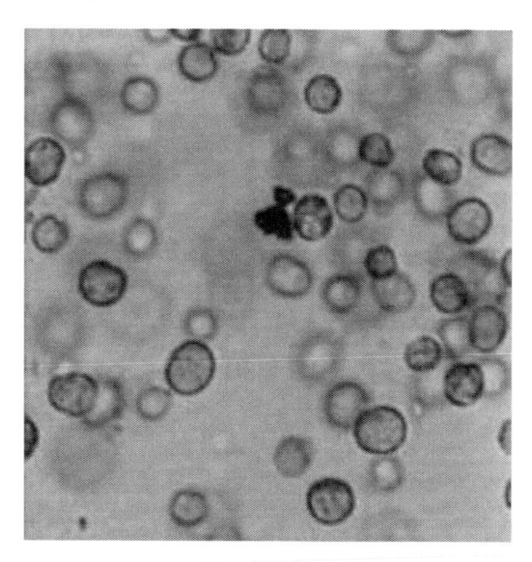

Figure 3. Zooxanthellae in the tissues of a coral polyp.

In shallow water, where light is abundant but wave stress is high, the colonies deposit robust branching skeletons; in deeper water, where light is scarce, the colonies form horizontal plate-like structures in which each polyp may harbor an increased number of *zooxanthelae*. Under highly adverse conditions such as prolonged darkness or freshwater flooding, it is no longer advantageous for the coral polyps to maintain their zooxanthelae and they expel them from their tissues. Since the skeletal-growth rate of corals is dependent on their algal partners, true reef-building corals are almost never found outside the range of stable symbiosis.

Some coral species harbor no zooxanthelae; some of these species are found in crevices under the large structures erected by reef-building corals. Many of them are solitary cup corals such as *Astrangia*, which encrusts shells and rocks as far north as Cape Cod. Such corals can tolerate lower salinities, lower temperatures and greater depths: up to 6,000 meters in the deep sea. Even the deep, cold waters of the Norwegian fjords harbor great banks of *Lophohelia*, a *colonial* branching coral. Although these nonsymbiotic corals are distributed worldwide, their rate of growth is much lower than that of their symbiotic relatives, and they do not form massive reefs. In isolated instances, colonies of these corals do contain symbiotic algae, but the algae do not appear to contribute significantly to the nutrition of their host.

The zooxanthelae are stored within individual membrane-bound cavities inside each of the cells in the stomach wall of the coral polyp. The feedback mechanism whereby the host regulates the number of its algal cells has not been determined, but there is little evidence that the corals “farm” and digest their algae. Instead, the coral polyps seem to control the population of zooxanthelae by extruding the older and less metabolically active algae. Robert Trench and his colleagues at the University of California at Santa Barbara have shown that specific strains of zooxanthelae are adapted to specific coral species. Some strains can live successfully in several different corals, and some corals are not discriminating about the lineage of their symbiotic algae. The fascinating problems presented by the symbiotic selectivity of corals are only beginning to be explored, and corals provide a valuable experimental system for the study of cellular interactions in general.

2.1.2. Soft Corals

Soft corals are soft bodies made up of a large number of polyps connected by fleshy tissue. They lack the limestone skeleton found on their relatives, the hard coral. The term 'soft' is a bit misleading because these corals have numerous tiny, needle-like spicules in their tissues (Figure 4, 5). Numerous sclerites are randomly scattered in the interior mezoglea layer, or cortex form directly below the ectodermal cell layer.

Apart from their swaying bodies and jelly like feel, soft corals are distinguished by the eight tentacles on each polyp and have a feathery appearance, whereas hard corals have smooth tentacles.

Soft corals may seem potentially more vulnerable to predators than those that have a stony skeleton, but, in reality, they are not. This is partly because of the presence of the spiky spicules, which function like thorns on a rose bush, and partly because soft corals contain powerful toxins (terpenes). Underwater, these toxins make the tissues of soft corals either distasteful or toxic to fish. They are also put to use in the constant battle for space. Soft corals introduce them into the water around them where they can kill neighboring hard corals and repel other soft corals. Soft corals are able to move, very slowly, by extending the tissues at their base. When their route crosses hard coral colonies, they kill the polyps, leaving a white, dead path behind them.

Figure 4. Soft coral *Umbellulifera* sp.

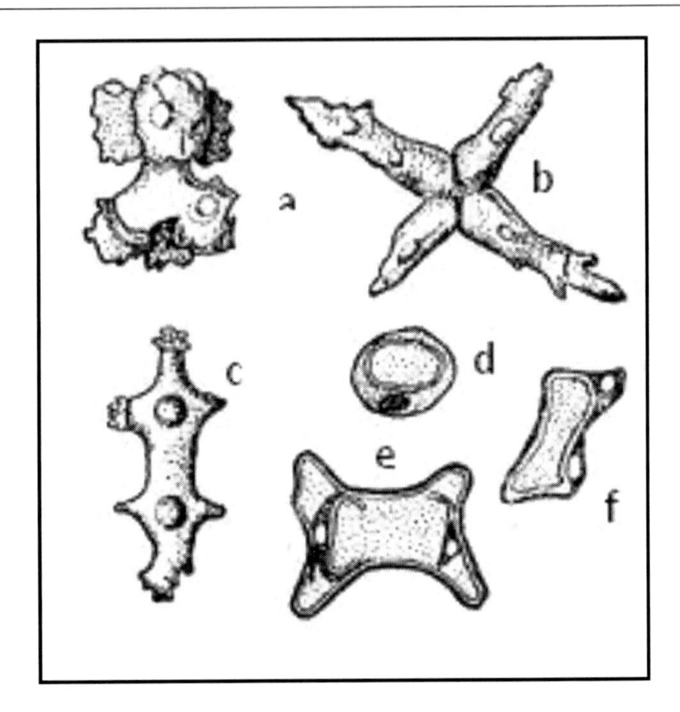

Figure 5. Appearance of spicules: a - Isishippuris - simple six-rayed spicule; b - Briareum arboreum- joint of 4 needles; c - Gorgonia fusco-purpurea - beam spicule; d-f - Alcyonium digitatum - three stages of development of the spicule.

Soft corals are much more likely to feed with their tentacles extended during the day than hard corals. Some contain zooxanthelae and appear brown in color while the bright color of the spicules is revealed in those without zooxanthelae. Only some species are able to retract their polyps. Some of these can also contract their entire structure when under stress such as in low tides.

Some corals produce hard skeletons but, because they have eight-tentacle polyps, are classified as soft corals. Another peculiarity of these species is the bright color of the skeleton that remains after their death. Organ pipe coral skeleton is red, but is obscured by polyps that feed much of the time. The tube-like skeleton resembles organ pipes and is actually composed of fused spicules, not solid limestone as in true hard corals.

It is important to treat corals well so they can be around for a very long time. Never remove or break off a piece of coral. When reef walking, never walk where you are not supposed to and stick to the sandy areas. When you are in a boat, never anchor near a coral reef because the anchor could be caught in the coral and destroy it.

2.2. The Physiology of Coral Symbiosis

The modern study of the physiology of coral symbiosis began with a series of elegant experiments done by C. M. Yonge on the Great Barrier Reef Expedition of 1929. Yonge showed that symbiotic corals take up phosphates and ammonia from the surrounding seawater by day and release them at night. In order to study this phenomenon in greater detail Thomas Goreau and Nora Goreau [15] supplied carbon in the form of the radioactive isotope carbon 14 to reef corals. During the daylight hours, the zooxanthelae assimilated the radioactively labeled carbon and photo synthetically fixed it into organic matter at a rate that was dependent on the intensity of the light. Some of this organic matter was then "leaked" from the algae to the coral host. Subsequent work by Trench and Leonard Muscatine of the University of California at Los Angeles and by David Smith of the University of Oxford showed that the leaked compounds include simple nutrients such as glycerol, glucose and amino acids. These compounds are utilized by the coral polyps in energy-yielding metabolic pathways or as building blocks in the manufacture of proteins, fats and carbohydrates.

It has long been known that the rates of metabolic reactions are strictly limited by the rates at which waste products are removed from the immediate environment. In higher animals, the task is accomplished by specialized circulatory and excretory systems. These systems are absent in the anatomically simple coelenterates, which rely largely on the slow process of diffusion to remove soluble inorganic waste products such as carbon dioxide, phosphates, nitrates, sulfates and ammonia. The zooxanthelae, however, need photosynthesis for the very substances the coral polyp must get rid of, and they are believed to actively take them from their host.

The *photosynthetic* demands of the zooxanthelae therefore result in the cycling of the coral's waste products into new organic matter. During the daylight hours, the symbiotic algae produce more oxygen than the coral polyp can utilize for its respiration, and some of the carbon dioxide produced by the respiratory process is refixed by the algae into new organic matter. Thomas J. Goreau, in order to estimate the efficiency of the internal carbon eyeling in corals, determined the abundance of carbon 13 in the coral tissue and skeleton. Carbon 13 is a rare but nonradioactive natural isotope, with respect to the abundance of the common natural isotope carbon 12.

For reasons not necessary to explain here, photosynthesis takes up carbon 12 slightly faster than it does carbon 13. Hence, the organic matter synthesized by the zooxanthelae will have a relative preponderance of carbon 12, and a pool of carbon compounds enriched in carbon 13 will be left behind. It is from the compounds in this pool that the calcium carbonate coral skeleton is built. By determining the relative amounts of the two isotopes with a mass spectrometer, it was estimated that about two-thirds of the carbon taken up in photosynthesis and calcification is recycled from the respiratory carbon dioxide of the coral polyp, with the rest being taken up from the seawater.

Organic matter leaked by zooxanthelae is only one of the three major sources of coral nutrition. Corals are efficient carnivores, immobilizing animal plankton with the stinging cells of their tentacles or trapping them on filaments of mucus that are then reingested. A polyp can detect a potential food item chemically, and it responds by extending its tentacles, by opening its mouth or by extruding its mesenterial filaments. James Porter of the University of Georgia has analyzed the content of the coral stomach and found that the polyps feed mostly on tiny crustaceans and worm-like plankton that hide in the interstices of the reef by day and emerge at sunrise and sunset.

Studies with radioactively labeled compounds have also shown that corals are able to take up dissolved organic matter across their body wall. Since corals actively feed on plankton, take up nutrients from seawater and absorb chemicals released by their zooxanthelae, they fill several *ecological* roles simultaneously: primary producer, primary consumer, detritus feeder and carnivore. This complex food web reduces their dependence on any single food source, which might be subject to random variation as environmental conditions change.

Two species of the coral *Agaricia* growing side by side differ strikingly in shape and size. One type has whorled fronds, whereas the other has shingle-like plates. Such complex morphological differences are produced by subtle environmental gradients, such as the decline of ambient-light intensity with depth. The corals are at a depth of 43 meters off the Jamaican coast.

2.3. Calcification in Corals

Growth in corals is achieved by an increase in the mass of the calcareous skeleton and the overlying living tissue. The skeleton of corals is composed entirely of *aragonite,* a fibrous crystalline form of calcium carbonate ($CaCO_3$); *calcite*, the commoner crystalline form of calcium carbonate, is absent (Figure 6). In the reef, many algae also deposit aragonite or a more soluble form of calcite with high magnesium content. Working in Bermuda, Heinz Lowenstam of the California Institute of Technology showed that some calcareous organisms tend to deposit the less soluble calcite in the cold seasons and the more soluble aragonite in the warm seasons, but the mechanisms by which organisms regulate the mineralogy of their skeleton are still unknown.

Coral polyps absorb calcium ions from seawater and transfer them by diffusion and by an active pumping mechanism to the site of calcification. Calcium ions are a major biochemical regulator of cell metabolism and must be kept at extremely low levels if the cells of a tissue are to function. Although coral tissues have a total calcium concentration similar to that of seawater, the concentration of free ions in them is much lower because most of the calcium is bound to membranes or to organic molecules, the calcium bound in these organic complexes turns over rapidly.

Nora Goreau, working in collaboration with Raymond Hayes of the Morehouse College School of Medicine in Atlanta, recently made detailed electron-microscope studies of coral polyps. In the course of these studies, minute calcium carbonate crystals enclosed within membrane-bound vesicles were observed in the outer cell layer of the polyp. The crystals are extruded through the membrane to the coral skeleton, where they act as nuclei for continued crystal growth. This work may serve to clarify basic mechanisms of calcification in the cells of a variety of organisms, particularly because corals lack the hormonal controls over calcification that complicate these mechanisms in more advanced organisms.

Figure 6. The appearance of the corallite skeleton.

The major obstacle to the study of the physiology of calcification in corals has been the difficulty of keeping corals alive and healthy in laboratory aquariums long enough to make accurate measurements of the calcium uptake. Thomas F. Goreau circumvented the problem by measuring calcification *in situ* in the living coral reef. This was done by providing the coral with calcium in the form of the radioactive isotope calcium 45 and measuring the uptake of the radioactive calcium into the coral skeleton. The method is so sensitive that growth can be detected in corals that have been exposed to the radioactive calcium for only a few hours, which is what makes field studies practicable.

Such studies have shown that although reef-building corals grow under uniform conditions of temperature, illumination and water circulation, there are very large differences in the growth rates of different species. The highest rates are invariably found in the branching corals, such as the West Indian elk-horn and staghorn corals. *Millepora* ("fire coral") is a close second, with the *Porites* ("finger corals") third. The massive corals grow more slowly. In the branching corals most of the growth takes place at the tips of the branches, and new branches develop almost anywhere on the older parts of the colony.

2.4. Reef Architecture

Coral polyps may not dominate the biomass (the total mass of living matter), the biological productivity or even the calcification in all parts of a coral reef. Nevertheless, the existence of many of the animal and plant communities of the reef is based on the ability of coral to build a massive wave-resistant structure. The dynamic interactions of the geological and biological processes that control the growth of coral reefs are well illustrated in the 150-mile fringing reef along the northern coast of Jamaica, which we have studied for the past 28 years.

The major structural feature of the living reef is a coral rampart that reaches almost to the surface of the water. It is made up of massive rounded coral heads and robust branching corals, which build a rigid, cavernous palisade of intergrown coral skeletons. Living on this framework are smaller and more fragile corals and large quantities of green and red calcareous algae. The biomass of these algae is small compared with that of corals, but their productivity and turnover are so high that the sand consisting of their skeletal remains makes up the bulk of the calcium carbonate deposited in the reef.

Hundreds of species of encrusting organisms live on top of the coral framework, binding the coral branches together with their thin growths.

Innumerable fishes and invertebrates also hide in the nooks and crannies of the reef, some of them emerging only at night. In addition, sessile organisms cover virtually all the available space on the underside of coral plates and on dead coral skeletons.

The crest of the reef runs parallel to the coast, in some places touching the shore and in others enclosing a sandy lagoon about five meters deep and up to a few hundred meters wide. This area is protected from the surf and is dotted with isolated coral heads. The lagoon is dominated by patches of calcareous algae and a *community* of bottom-living animals, notably sea urchins and sea cucumbers, which earn their keep by filtering organic matter out of the sediments or the overlying water. Many of these organisms graze on filamentous algae; if the grazers are removed from an area of the lagoon, a dense mat of algae forms after only a few days. The burrowing and churning activities of the grazers are important because they release nutrients created by the bacterial decomposition of organic matter buried in the sediments. Dense "lawns" of the sea grass *Thalassia* form special habitats harboring their own community of sea urchins, conches and many other species.

Seaward of the reef crest is the fore reef, where corals blanket nearly the entire sea floor (Figure 7). The corals form massive buttresses separated by narrow sandy channels, down which passes a steady flow of fine sediment originating with the disintegration of dead corals, calcareous algae and other organisms. The channels resemble narrow winding canyons with vertical walls of solid coral growth. They may be as much as 10 meters deep, and some are completely roofed over with coral. This dramatic interdigitation of buttresses and channels dissipates wave energy and at the same time allows the free flow of sediments that would otherwise choke the growth of the coral.

Down from the buttress zone is a coral terrace, a slope of sand with isolated coral pinnacles, then another terrace and finally an almost vertical wall dropping into the darkness of the greater depths. The distribution of coral species and other animal communities in the reef are zoned by depth, a feature that enables paleontologists to study a section of an ancient reef now on dry land to accurately estimate the original depth of that section from the fossil animals associated with it. In water deeper than 100 meters, few algae or symbiotic corals grow well because of the low light levels and the fauna is dominated by animals that catch or filter the organic detritus sifting down from the reef above. The detritus feeders include the true sponges, the antipatharians ("precious corals") and the gorgonians (sea fans). In addition, common here are the sclerosponges, an ancient group that were major reef builders in the geological past but were long thought to have become extinct hundreds of millions of years ago. Our diving studies of the deep reefs of Jamaica showed them to be alive and well but displaced by the deeper habitats.

Figure 7. Algal-coral zone at Vietnamese reef.

2.5. Reef Growth

The growth of the reef is the result of a dynamic relation between the upward extension of the coral framework and the flushing away of a much larger volume of fine-grained detritus. The export of sediment from the reef is largely accomplished by gravitational flow and creep, either into the lagoon or down the channels of the buttress zone into deep water. Unstable piles of coral may also grow until they topple under their own weight and slide away. When the lower Jamaican reef was explored in the research submarine Nekton Gamma 11 at depths of more than 200 meters, enormous piles of sediment and huge blocks of solid reef were observed at the base of the drop-off; they may have been dislodged by earthquakes. Such dislocations create fresh substrates for encrusting organisms and help to establish coral communities on the steep lower slopes, particularly the plate-like whorled colonies of *Agaricia.*

Two other major processes influence the growth of the reef: biological erosion and submarine lithification. Many species of filamentous algae, fungi, sponges, sea worms, crustaceans and mollusks bore into coral skeletons, excavating holes by mechanical rasping or chemical dissolution. The commonest is the boring sponge *Cliona,* which saws out tiny chips of calcium carbonate; the chips are a major component of the fine sediments. *Cliona* can riddle a coral skeleton with holes without damaging the living coral polyps. In the deeper waters, many corals grow in flat, thin sheets to maximize their light-gathering area and hence are quite susceptible to erosion by borers, which can cause the corals to break off and fall down the slope. In some places, however, the coral is so overgrown with encrusting organisms that it remains in place even though it is no longer directly attached to the reef.

Counteracting the effects of biological erosion is submarine lithification: the deposition of fine-grained carbonate cement in the pores and cavities of the coral skeleton. Sediments trapped in the reef framework are rapidly bound together by encrusting organisms and the calcareous cement. The origin of the cement is not yet clear; it may be an inorganic precipitate manufactured by bacteria that live in the crevices of the reef. Studies at the Discovery Bay Marine Laboratory in Jamaica, done in conjunction with Lynton Land of the University of Texas at Austin, showed that once the cement has hardened it is in turn bored and refilled; the filled holes are apparent when thin sections of the aggregate are examined under the microscope.

Submarine lithification results in the outward accretion of the fore reef and stabilizes the steep profile of the drop-off wall. The growth of reefs is therefore the product of a dynamic balance among framework growth, sediment transport, bio erosion by borers, mechanical destruction and submarine lithification, with the relative importance of these factors varying from reef to reef (Figure 8, 9).

The living reef is basically a veneer growing a few millimeters a year on top of a complex topography of superposed ancestral reefs. In Jamaica as much as nine meters of reef has built up since the present sea level stabilized some 5,000 years ago. The ancient reefs remain, providing a record of changes in sea level and of the uplift of land by the movements of tectonic plates.

The rise and fall of the sea level over the past few million years has been caused by changes in the volume of water tied up in land glaciers and ice sheets during the Pleistocene ice ages. When ice sheets grew in the Northern Hemisphere, the sea level dropped and coral reefs were stranded above the waterline.

Figure 8. An erosion of colonies Astreopora and Porites by polychetes and mollusk. The arrow indicates the border aggression colonies of different species.

Figure 9. Forming of framework by various animals at recent reef.

Today fossil ridges and wave-cut notches mark the ancient sea level. A succession of stranded reefs are found in Jamaica, Barbados, New Guinea and on other coral coasts; these reefs were formed 80,000, 105,000, 125,000 and

200,000 years ago, when the climate was warmer and the sea level higher than it is today. Conversely, in Jamaica a series of drowned and overgrown ridges can be seen at 25, 40 and 60 meters below the present sea level. These drowned reefs were formed during periods of intensive glaciation 8,000, 11,000 and 14,000 years ago, when the sea level was considerably lower than it is today. The ancient reef is therefore a dimly visible palimpsest under the living reef, like a medieval manuscript that has been repeatedly erased and written over but shows faint traces of its history. Such features help in establishing the chronology of the Pleistocene ice ages and the volume of water added to the oceans by the melting of the ice.

2.6. Reef Ecology

The history of the modern Jamaican reef since the sea stabilized at its present level 5,000 years ago has not been long enough to establish a community: an ecosystem in equilibrium. This fact is evident from the almost haphazard development of reefs along any coral coast: some areas have well-developed reefs and others have only isolated patches of coral. Often there are no obvious environmental influences or catastrophic factors (such as earthquakes or tidal waves) that would explain such differences in development. It seems rather that chance variations in the settlement of free-swimming coral larvae and growth play a major role in determining the formation of reefs, and that there simply has not been enough time for corals to occupy all favorable habitats.

The role of chance in coral settlement is also reflected in the variability of the major species that fill the same structural roles in any reef. In some Vietnamese reefs the dominant coral is the branching coral *Porites cylindrica,* but in similar habitats the same role is filled by the different species *Montipora digitata,* which forms colonies of identical shape, size and orientation. Hence, the creation of diversity in a coral reef historical variation is in many reefs just as significant as the approach to an ecological equilibrium where many specialized organisms coexist.

The many localized habitats and species in the reef give the reef community a wealth of interactions within and among species whose complexity can only be dimly grasped. An intuitive understanding of the major interactions can be gained only after years of field experience. Even then, one may focus on so few components of the community that it is easy to miss the significant roles played by many obscure, unexamined or unknown organisms.

The intense competition for food and space in the reef habitat has given rise to a wide variety of survival strategies. Corals growing close together compete for space, and some species are able to extrude mesenterial filaments from the gut to kill the polyps of adjacent colonies. Among coral species there is a hierarchy of aggression such that slow-growing but aggressive corals can avoid being overgrown by faster-growing but less aggressive ones. This process may lead to an increased diversity of species. In some instances, however, the result is precisely the opposite: James Porter has found that in the reefs on the Pacific coast of Panama the overwhelmingly dominant coral, *Pocillopora damicornis,* is both the fastest-growing and the most aggressive.

Grazing on algal and coral tissues by fishes, sea urchins and other animals has two important effects. Selective grazing may keep a few dominant species of algae from crowding out the more marginal species, so that a diversity of species is able to exist. Experiments in which grazers are excluded from an area of the reef usually result in choking densities of a few dominant algal species, which is rare under ordinary circumstances. Grazers that scrape tissues of hard substrates also create fresh surfaces where new algae can grow and the larvae of sessile organisms can settle. Some fish species systematically kill patches of coral tissue so that "farms" of algae can grow on the bare coral skeleton (Figure 10). The fishes, which graze on the algae, chase any intruders on their territory, including much larger fishes and even human divers. How much damage to the reef is done by such biological space clearing compared with that done by slumping and storms is not known.

The coral colonies are still varied but smaller in size, and much of the available space is occupied by sand-producing calcareous algae, sponges and large gorgonians (sea fans). The deep fore reef extends from 30 meters to 70. This zone has a steep topography and is poorly illuminated, with a light flux about 5 percent that at the surface. At depths below 30 meters coral growth becomes patchy, with a progressive reduction in number of species and size and density of colonies. There is also extensive transport of sediment from the shallow zones above. Beyond the deep fore reef the vertical wall drops off into darkness.

Much remains to be learned about the nutrient and energy cycles of reefs. The richness of reef biological processes in the face of the poverty of dissolved nutrients in tropical surface waters is evidence that there is an efficient internal cycling of nutrients within the reef ecosystem, but the matter has yet to be investigated in enough detail. The major limiting nutrient in the oceans is generally thought to be nitrogen, and in coral reefs large amounts of the atmospheric nitrogen dissolved in the seawater are fixed in utilizable forms by filamentous blue-green algae. Another source of nitrates is the oxidation of ammonia by bacteria in the course of the decomposition of organic matter in the sediments of the reef lagoon. Recent work indicates that the oxidation of ammonia to nitrate is particularly intense in the fine-grained organic sediments trapped by the roots of sea-grass beds.

Figure 10. Aggression algae on damaged corals.

The coral reefs of the Atlantic, the Caribbean and the Indo-Pacific do not differ fundamentally in their structural forms, their habitats and the interactions of their species, even though the organisms occupying specific ecological roles vary greatly between oceans and even between individual reefs. Between the Pacific and the Caribbean, however, there is one significant difference: in the Pacific the active growth of coral only goes down to 60 meters, and in the Caribbean it goes down to 100 meters. The reduced range in depth of the Pacific corals may be due in part to periodic infestations by the crown-of-thorns starfish *(Acanthaster planci),* which feeds on coral by turning its stomach inside out, spreading it over the coral and digesting the coral tissues (Figure 11).

Before the recent well-publicized outbreak of *Acanthaster,* the organism was limited to deeper water and was rarely seen. Then an unexplained population explosion gave rise to a food shortage that forced the starfishes to move up to shallower water, where their destructive effects were readily apparent. The lower limit of reef growth in the Pacific may therefore be affected by periodic starfish grazing. Much remains to be done to prove the hypothesis, however, not least because many Pacific reefs also show signs of being more intensively eroded mechanically than the Caribbean reefs.

Figure 11. Acanthaster planci eating corals.

These points illustrate some of the handicaps ecologists face in attempting to predict the stability of reef populations in response to environmental changes or the sensitivity of reef food networks to alterations in the abundance of particular species. Since coral reefs are localized centers of high biological productivity and their colorful fishes are a major source of food in tropical areas, many marine biologists view with alarm the spread of tourist resorts along coral coasts in many parts of the world. Such developments are almost always accompanied by increased dumping of sewage, by over fishing, by physical damage to the reef resulting from construction, dredging, dumping and landfills, and by destruction of the reef on a large scale to provide tourists with souvenirs and coffee-table curios. In many areas (such as Bermuda, the U.S. Virgin Islands, Hawaii and Vietnam), development and sewage outfalls have led to extensive eutrophication: the overgrowth and killing of the reef by thick mats of filamentous algae, which in turn support the growth of oxygen-consuming bacteria. The results, which are being intensively studied by the author of this book and his colleagues at the A.V. Zhirmunsky Institute of

Marine Biology [30, 34, 36, 37], include an increased sensitivity of corals to bacterial diseases, the death of living coral and the resulting erosion of the reef, and the generation of foul-smelling hydrogen sulfide.

2.7. Habitat and Distribution

Various species of corals are found in all oceans of the world, from the tropics to the Polar Regions. Reef-building corals are scattered throughout the tropical and subtropical Western Atlantic and Indo-Pacific oceans, generally within 30°N and 30°S latitudes.

Western Atlantic reefs include these areas: Bermuda, the Bahamas, the Caribbean Islands, Belize, Florida, and the Gulf of Mexico.

The Indo-Pacific ocean region extends from the Red Sea and the Persian Gulf through the Indian and Pacific oceans to the western coast of Panama. Corals grow on rocky outcrops in some areas of the Gulf of California.

Although various types of corals can be found from the water's surface to depths of 19,700 ft. (6,000 m), reef-building corals are generally found at depths of less than 150 ft. (46 m), where sunlight penetrates. Because reef-building corals have a symbiotic relationship with a type of microscopic algae, sunlight is necessary for these corals to thrive and grow. Reefs tend to grow faster in clear water. Clear water allows light to reach the symbiotic algae living within the coral polyp's tissue. Many scientists believe that the algae, called zooxanthelae, promote polyp calcification. See adaptations for more information on this algae and its relationship with coral. Light-absorbing adaptations enable some reef-building corals to live in dim blue light. Reef-building corals require warm ocean temperatures (68° to 82°F, or 20° to 28°C). Warm water flows along the eastern shores of major landmasses.

Reef development is generally more abundant in areas that are subject to strong wave action. Waves carry food, nutrients, and oxygen to the reef; distribute coral larvae; and prevent sediment from settling on the coral reef.

Precipitation of calcium from the water is necessary to form a coral polyp's skeleton. This precipitation occurs when water temperature and salinity are high and carbon dioxide concentrations are low. These conditions are typical of shallow, warm tropical waters. Most corals grow on a hard substrate.

Chapter 3

CHARACTERISTICS OF REEF

3.1. GEOLOGICAL HISTORY AND FORMATION

The earliest reefs developed 2 billion years ago, during the Pre-Cambrian era. These reefs were built by colonies of calcareous algae, not coral. Corals have been present in the warm seawaters of the world for over 500 million years, contributing to the formation of reefs all around the world (Figure 3.1).

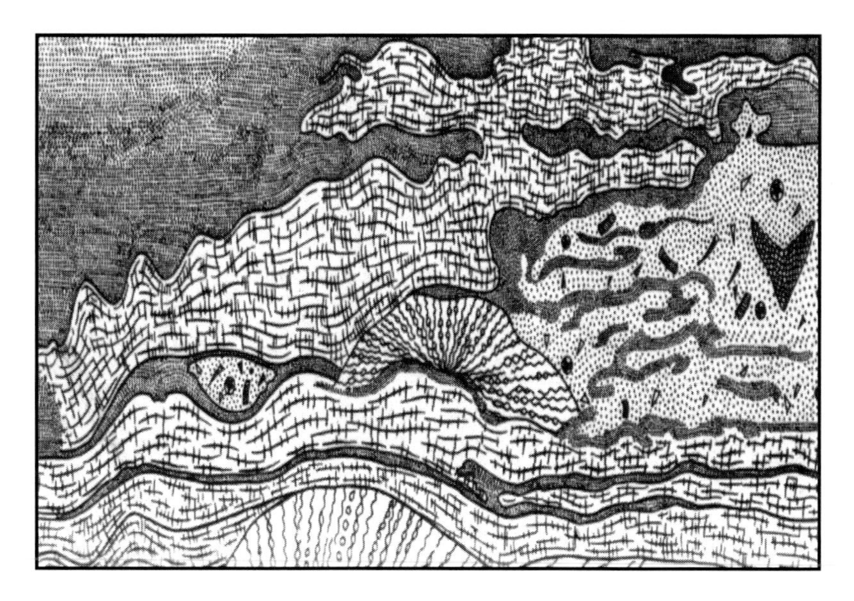

Figure 3.1. Regeneration of a part of fossil Paleozoic reef after covering its soft deposit.

Early in the Earth's evolutionary history, corals lived in the Tethys Sea. This body of water was vast, circling the globe with a body consisting of the Pacific, central Atlantic, Mediterranean Sea, and Indian Oceans. The water fauna was continuous with no distinct bio-geological zones such as they exist today.

Corals became separated geographically and evolved into new species. Other events during the previous one million years aided in the adaptation of corals to those changes. Ocean currents, ice ages, and fluctuation in sea levels have been linked to the phenomenon of the species we currently see in our coral reefs. It is currently under study that the corals of today bare little to no resemblance to those of the Paleozoic era and time previous to that. As recent as the early Triassic period, the oceans were anoxic, increased levels of carbon dioxide, and global shifts of ocean chemistry that reduced carbonate deposition. The fossils found since the mid-Triassic period show corals with aragonite calcification, species such as we see today.

Even the number of coral species that evolved during the Triassic period was nearly wiped out by the upheaval of tectonic plate movement during the Miocene period 25 million years ago and once again during the ice ages. Around 15,000 years ago, the seas were about 120 meters lower than they are now, and 7,000 years ago, they were about 20 meters below the present level. Estimates show that modern reef growth, as related to present sea level, must have begun about 5,000 years ago [20]. It also appears through fossil records that modern reefs are only patchy in comparison with the flourishing reefs prior to the Pleistocene epoch.

Charles Darwin's theory of coral formation recognizes three types of reefs: the fringing reef, the barrier reef, and the atoll. Fringing reefs border shorelines of continents and islands in tropical seas. There will be a narrow channel of water separating the coral from the land. In these shallow waters, coral communities begin their formation by growing upward toward the water's surface and then horizontally out to sea. As the land sinks slowly due to plate movement, the corals continue to grow upward. If the land sinks faster than the coral can grow, so that the coral sinks below 46 meters, the colony dies from lack of sunlight.

The next type is the barrier reef, which occurs farther offshore. Barrier reefs form when landmasses sink, and fringing reefs become separated from land by wide channels. Landmasses sink as a result of erosion and shifting (uplift or submergence) tectonic plates of the earth.

The last stage is the coral atoll. The landmass is small - such as avolcano - it may eventually disappear below the ocean surface, and the reef becomes an atoll. Atolls are reefs that surround a central lagoon. The result is several low coral islands around a lagoon. Atolls commonly occur in the Indo-Pacific. The largest atoll, named Kwajalein, surrounds a lagoon over 97 kilometers long.

At one time it was mistakenly thought that coral grew at the bottom of deep tropical seas and succeeding generations grew on top of the dead calcium carbonate skeletons. With all three of Darwin's reef formations, his theory is based on the premise that all corals' original substrate was volcanic basalt. In 1952, on the Eniwetok Atoll in the Marshall Islands, Darwin's theory was proven correct. After drilling through 4,630 feet of limestone, the U.S. Geological Survey hit basement rock - olivine basalt. They had dug back through over 5 million years of history. Not all atolls and reefs are based on that much history, but others have even older stories to tell [9].

Figure 3.2. View of the atoll from the top.

3.2. Reef Composition and Biota of the Reef

3.2.1. Plant Communities

Bottom vegetation, coral reefs may at first glance seem to be lean compared with the lush thickets of macrophytes and sea grasses in the coastal zones of the seas of the temperate zone. Indeed, the flora of aquatic vegetation in temperate waters includes species several times richer than flora inhabiting coral reef *biotopes*. There are no large thallus Macrophytes such as kelp. Macrophytes with thallus more than 1 m on reefs are rare. Most of them are very small forms with long threadlike 3-6 cm thallus.

The composition of the aquatic vegetation reef has (up to 30-40% biomass) massive cortical calcareous coralline algae. *Benthic* primary production plant communities submitted to the dense thickets of macrophytes and cortical coralline, which accounted for 40-80% of the total primary production of demersal reef environments. On reefs, where living corals are few, calcareous algae dominate the benthic biotopes and play a leading role in the production of organic matter and carbonate material.

The results of the analysis of the cores obtained during drilling of reefs indicate that carbonate material produced calcareous algae, which can be up to half or more of the total number of calcareous material. A large proportion of Red limestone algae and specialized green calcareous algae from the genus Halimeda, consisting of seaweeds, are the most salient features of the bottom of the vegetation's reef. Bottom vegetation of the reef is exposed to intense consumption of many fish and benthic invertebrates. This process is one of the main factors governing the composition of benthic habitats and aquatic vegetation on the reef. Thanks to the grazing algae, it continuously rejuvenates their populations with the dominance of fast-growing small thallus species. In combination with high specific surface areas occupied by aquatic vegetation, it provides significant production in benthic habitats.

The plant grouping reefs are composed of representatives of the four major taxonomic groups of diatoms, algae, blue-green, green, Brown and red, as well as a group of flowering aquatic plants-grasses marine (Figure 3.2.1).

The most common reef algae are representatives of a specialized species of reef ecosystems called *Halimeda*. This calcareous alga has heavily inlaid calcium carbonate from the highly durable articulate thallus. They are attached to the rhizoid and friable rocky substrates. Halimeda is often viewed as bottom vegetation of the reef, such as calm lagoons, rocky *reef flats*, and in the *reef slopes* down to its foundation. One reef could have 10-20 species of *Halimeda*. Alga substrate surface coating can reach 60-80% and *biomass* up to 20 kg/m^2.

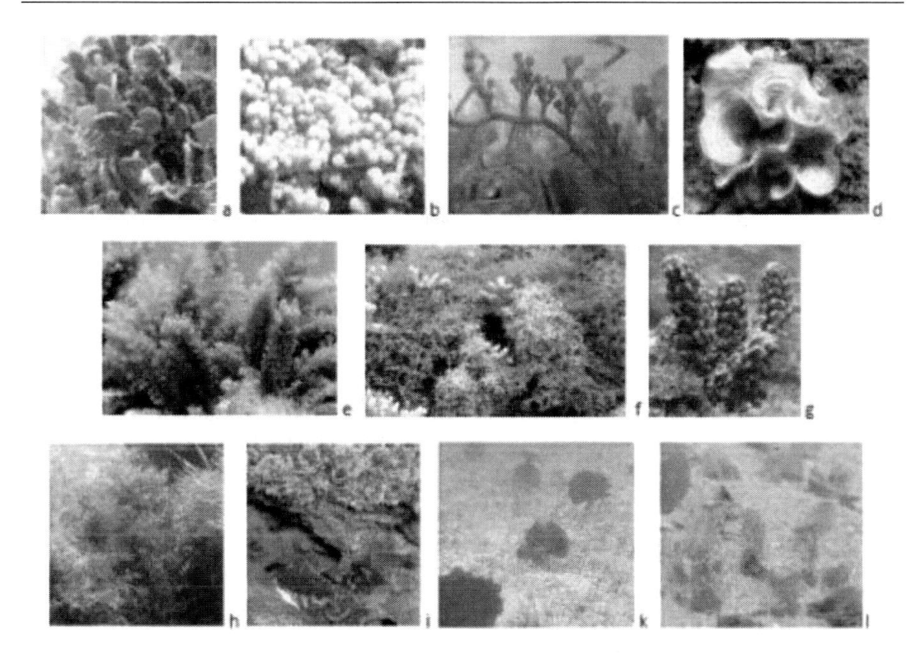

Figure 3.2.1. Different species of seaweed: a- Halimeda discoidea, b- Caulerpa racemosa, c- C. lamourouxii, d- Padina australis, e- Asparagopsis taxiformis, f- Chnoospora implexa, g- Turbinaria ornata, h- Sargassum quinhonese, i- Litotamnium sp., k- Avrainviellea sp., l- Halophila ovalis.

Green algae are usually composed of epiphyte and periphyton. They play a prominent role in the early stages of succession in reef communities. Green algae *Ulva, Enteromorpha, Caulerpa, Cladophora* predominate among the first settlers in the recovery stages of plant associations after typhoons and hurricanes. Intensive algae, Wintergreen bilberry species *Vaccinium myrtillus,* many fishes and the first settlers of invertebrates largely restricted their development. However, they are saved at the expense of species with high growth rate. Having the original openwork form, lichens such as algae *Caulerpa, Padina* are often found in shallow waters of many reefs and are therefore easy to remember, which really helps in the study of algal reef communities.

The total number of kelp on the reefs of the open ocean is usually low, anywhere from 10 to 30 species. However, many of them are often dominant in the composition of plant communities and are among the last representatives of the genera *Turbinaria, Dictyota, Sargassum.* In general, they gravitate to the intertidal, reef, lagoon fletu patch reefs with relatively high water column turbulence. Some types of *Sargassum* can create dense monospecific settlements, such as in the Indo-Pacific and Atlantic reefs. In the coastal zone of the many reefs of Vietnam, especially in its southern part, one or two *Sargassum* forms a solid thicket with the degree of substrate covering 40-70% with biomass from 4.5 to 6 kg/m^2. Some of the kelp-*Padina, Turbinaria* are found on reef slopes.

Corallines algae, *Porolithon, Sporolithon, Litophyllum* and similar algae play a special role in formation of the windward reefs, particularly reef flats.

Their massive, encrusting, weakly dissected or blow-like colonies are very firmly attached to the substrate and provide wave stress. Some characteristic types of corallines definitely occupy *ecological niches*, having the appropriate range of environmental adaptation. Some of them can withstand the strongest direct solar radiation and long drying, others are resistant to direct wave effects. Thanks to the properties of the corallinae, they are the main reef-builders in the wave area of the reef-front (Figure 3.2.2). Other types prefer more shade and are less resistant to water stress under live colonies of corals and macroalgae, on the walls of the channel. Maximum dissemination falls on the Leeward side and the reef flat on the slope of the reef, at depths of more than 10 m. Corallines and especially *Poroliton* transferred periodic drying better than thallus macrophytes, which can fully occupy the elevated flat plots, creating a breakwater and the crest of the reef.

Flat plots, where there is massive eating of macrophytes fish and sea urchins, occupy the coralline algae because the latter is only available for some fish-parrots.

This was shown in the experiment. The reef flats have established cells, impeding the grazing algae fish. Thallus macrophytes quickly occupy protected sites, whereas open areas are dominated by corallines. Finally, corallines are better than corals, and can withstand pollution, replacing them on the reef areas exposed to anthropogenic effects.

Sea grasses and coral reefs prefer shallow coastal areas, lagoons and reef slope platforms held by coral sand or silt deposition. On such sites, they form dense thickets of type meadows (sea grass beds). These thickets are predominantly of one or two species, one of which usually dominates. Therefore, the Vietnamese reefs in shallow water formed a solid thicket of two species of Sargassum for 30-50 meters with 40-70% of coating and biomass up to 6 kg/m^2. The reef slope platform is often found in less dense meadows of white herb *Halophila.* Peculiar glades-seabeds are very characteristic in meadows of sea grass in shallow waters. These plots are formed because of the fish and urchins. During the day, they hide under rocks and coral reef patch and consume grass, near their safe havens.

Bottom vegetation has important functional value in the ecosystem of the reef. Bottom reef plants produce a significant proportion of the total autotroph of organic substances. Their contribution to the overall production of photosynthesis is estimated at 30-50%. An important feature of benthic plant associations of the reef is that they have small thallus or filamentous forms. They are also herbs that are readily eaten by many fishes and invertebrates. The direct use of a large part of primary products produced by bottom flora reef, shortens the food chain and improves the energy efficiency of the ecosystem of the reef. Bottom reef plants serve as a direct power source for approximately a quarter of all fish species that live on reefs.

Figure 3.2.2. Reef front, overgrown by various cortical limestone algae (by: Veron, 1986).

3.2.3. Mangroves

One of the most important components in plant associations on reefs are mangroves. Coastal mangroves are very widespread in the tropics, occupying about 70% of the entire coastline. Mangrove trees belong to different families, adapting to life in places that are filled with seawater. Mangroves, these evergreen deciduous forests in bays, lagoons and estuaries, are trees of 5-15 m with many branched support and aerial roots, and grow up to two meters above the ground. It is mostly an arboreal species, but they differ in their features. Long stem roots, as in ordinary trees, are unsuitable here in mud, where surface oxygen near the surface is absent, and poisonous hydrogen sulphide accumulates. Therefore, mangrove trees have peculiar roots (Figure 3.2.3).

Leaves are high above the ground, and ground breathing roots stick out of the soil and crankshafts. All these tweaks enable mangroves to provide air plants and therefore oxygen. Because the new IL constantly settles, the roots of the mangroves need to grow in parallel with increasing silt deposits. Mangrove vegetation needs special devices to help populate new mudflats shoals.

a b

Figure 3.2.3. The roots of the mangrove trees (a) and their fruit (b).

Since seeds are surrounded by water or viscous sludge, simple seeds will not be able to sprout in such circumstances. Many mangroves seedlings develop from flowers or fruits directly on the tree in the form of long sharp rods. They come off the tree and stick in the mud, or they become fixed when large amounts of water are transported to another place. However, representatives of various kinds have independent similarities. Mangroves occupy a limited area between the lowest sea level during low tide and the highest tide. This area is filled 10-15 times per month, but not daily, so the period during which it is under water is about 40%.

3.2.4. Fish and Zoobenthos Communities

The coral reef ecosystem is a diverse collection of species that interact with each other and the physical environment. The sun is the initial source of energy for this ecosystem. Through photosynthesis, phytoplankton, algae, and other plants convert light energy into chemical energy. As animals eat plants or other animals, a portion of this energy is passed on. Sponges have been a part of the coral reef ecosystem from early on. Several species of these porous animals inhabit reefs. Several hundred of gallons of water per day may pass through a sponge in order for it to obtain food and oxygen. Sponges provide shelter for fishes, shrimps, crabs, and other small animals. They appear in a variety of shapes and colors. Although the bright colors advertise that they are sponges, unpleasant secretions and noxious odors provide some protection for many species.

Sea anemones are close relatives of corals, without the stony skeleton. Anemones secrete a slimy, sticky substance that enables them to secure a footing. When a new site is desired, it peels itself off and moves to its new location. Indo-Pacific reef anemones are known for their symbiotic relationships with clown fish and anemone fishes. Clown fish are protected from the anemone's stinging tentacles.

In fact, the clown fish rarely strays more than a few feet from its host. An anemone's tentacles provide refuge for these fishes and their eggs. In return, anemone fishes may protect the anemone from predators, such as butterfly fishes. Anemones may even remove parasites from their host anemones.

Bryozoans encrust the reef. These microscopic invertebrates form *branching colonies* over coral skeletons and reef debris, cementing the reef structure. The reef is also home to a variety of worms, including both flatworms and polychetes. Flatworms live in crevices in the reef. Some *polychaetes,* such as Christmas tree worms and feather duster worms, bore into coral skeletons. Sea stars, sea cucumbers, and sea urchins live on the reef. The crown-of-thorns sea star is a well-known predator of coral *polyps.* Large numbers of these sea stars can devastate reefs, leaving behind only the calcium carbonate skeletons. In dead reefs, recently killed by the crown-of-thorns sea star, larger food and game fish are almost totally absent. Even deep-sea fish populations may be affected by this breakdown in the food chain.

Octopuses, squids, clams, scallops, nudibranchs (sea slugs), and marine snails are all mollusks that live on or near the reef. These shell-less mollusks appear defenseless, but they have evolved in creative ways to repel attackers. Some nudibranchs swallow the stinging cells of sea anemones whole. They store them, undigested and "unfired," in the gills on their backs. Later, they can use them for their protection. Other nudibranchs ooze toxic chemicals (sulfuric acid or poisonous slime) from their skin. An octopus uses a squirt of ink to cloud the water while it quickly jets away from its attacker. Carnivorous snails are capable of drilling holes into clams or other shelled animals and then eat them. One of the largest mollusks on the reef is the giant clam - capable of reaching a length of 1.2 meters.

Figure 3.2.4. Fish clown.

Both schooling and solitary fishes are essential residents of the reef ecosystem. Fish play a vital role in the reef's food web, acting as both predators and prey. Their leftover food scraps and wastes provide food and nutrients for other reef inhabitants. Some species of sharks, skates, and rays live on or near the reef (Figure 3.2.4). Others swim in to eat. Shark species include lemon, nurse, Pacific black tip shark, white-tipped reef, and zebra sharks. These sharks, as well as rays, generally eat crabs, shrimps, squids, clams, and small fishes. Parrotfish use chisel-like teeth to nibble on hard corals. These fish are herbivores and eat the algae within the coral. They grind the coral's exoskeleton to get to the algae and then defecate sand. A single parrot-fish can produce about five tons of sand per year. Wrasses comprise a large group of colorful cigar-shaped fishes.

Some species are known as cleaners, and set up cleaning stations along the reef. When a larger fish aligns itself at one of these cleaning stations, a cleaner wrasse removes parasites from the fish. If the same two fish met anywhere else, the larger fish would eat the smaller one.

Figure 3.2.4.1 Schools of fish on the reef slope.

However, it appears that different rules apply at the cleaning stations. Other fishes found on the reefs include an incredibly diverse group: angelfishes, butterfly fishes, damselfishes, triggerfishes, seahorses, snappers, groupers, barracudas, and puffer-fishes. Angelfish have developed a taste for only a certain kind of sponge or seaweed. An individual angelfish hovers close to its food source and uses its bright warning colors to defend it from all others interested in similar food, shelter, or nesting sites. Seahorses eat continuously, as they have no stomach to store food, forcing them to eat constantly. A young seahorse may eat as many as 3,500 shrimp in a day. Groupers mature first as females and produce eggs. They change sex later in life to become functioning males.

Eels are one of the reefs top predators. These fish live in crevices in the reef and venture out at night to hunt and feed. They have sharp teeth set in a powerful jaw. Eels eat small fishes, octopuses, shrimps, and crabs. Sea snakes are rarely found on reefs but do inhabit the waters around reefs in the Indo-Pacific. They possess small fangs but inject potent venom.

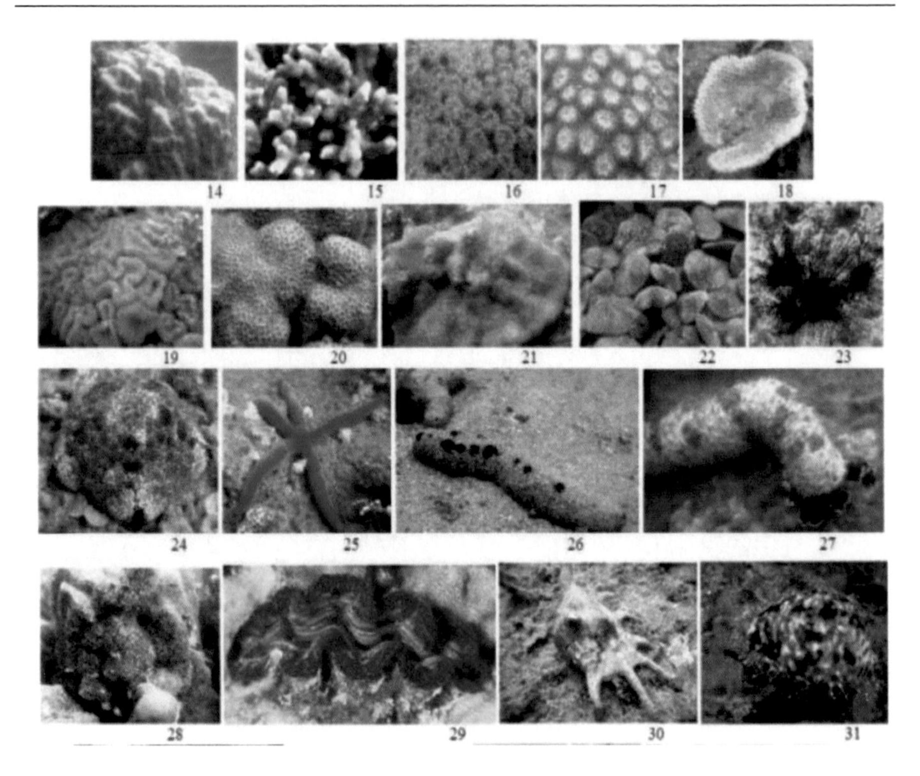

Figure 3.2.4.2. A common species of reef biota: 14-21 – colonial corals, 22 – mushroom corals, 23 - sea urchin, 24-25 – sea stars, 26-27 – holothurian, Spirobrachus giganteus, 29 – Tridacna crocea, 30 - Lambis lambis, 31 - Cyprea tygris.

Some sea turtles frequent reef areas. Green, loggerhead, and hawksbill sea turtles live in the warm waters of the Great Barrier Reef. These turtles will migrate thousands of miles in order to return to the same beach of her birth. She digs a hole in the moist sand to cradle hundreds of slippery, Ping-Pong ball-sized eggs. After carefully covering and packing the hole, she shovels dirt loosely all around with her front flippers to hide the nest from predators. Once the eggs hatch, the baby turtles face a gauntlet of seabirds, crabs, and lizards as they scramble across the sand to the sea.

3.3. Reef Formation and Types of Reef

The larvae of coral organisms attach themselves to hard substrates (rocks or previously installed madreporarian). Those which survive, the numbers of which are variable, initiate the colonization of the substrate, along with a whole set of associated symbiotic fauna. This phase takes place in shallow waters, where the coral skeletons accumulate quickly and the calcareous algae, which serve as an adhesive cementing the components together, also thrive. The detritus (foraminifera, *mollusks*, urchin spines, coral debris), which collects in the interstices, serves to solidify the overall structure. The whole complex is constantly being re-processed.

Coral reefs were classified in terms of their morphology by Darwin, as follows: fringing reefs, barrier reefs, flat reefs and atolls.

Fringing reefs lie near emergent land (Figure 3.3.1). They are fairly narrow and recently formed. They can be separated from the coast by a navigable channel, (which is sometimes incorrectly termed a "lagoon").

Barrier reefs are broader and lie farther away from the coast (Figure 3.3.2). They are separated from the coast by a stretch of water, which can be up to several miles wide and several tens of meters deep. Sandy islands covered with a characteristic pattern of vegetation sometimes form on top of a barrier reef. The coastline of these islands is broken by passes, which occupy the beds of former rivers.

Atolls are large, ring-shaped reefs lying off the coast, with a lagoon in their middle. The emergent part of the reef is often covered with accumulated sediments and the most characteristic vegetation growing on these reefs are coconut trees (Figure 3.3.3).

3.4. Distribution of Coral on the World

3.4.1. Generic versus Species Contours

Generic contours, familiar to bio geographers, tend to be accepted as being representative of the real world. But in this they are flawed, the worst flaw being that all genera have equal weight. This has no major effect on Atlantic diversity contours, where genera are poorly spectated, but in the Indo-Pacific, where the mean number of species per genus is 8.8, the distortion is major. At extremes, *Acropora,* with about 150 species, contributes the same amount of information as each of 40 genera that have only one species.

Figure 3.3.1. Fringing reefs.

Figure 3.3.2. Barrier reefs.

Figure 3.3.3. Atoll.

A second flaw is that generic contours are distorted by higher-level taxonomic boundaries; thus, the family Acroporidae, with approximately 250 species, contributes 4 genera, while the family Faviidae has 133 species. At present, these flaws cannot be redressed by creating a contour map of species distributions. The quantity and quality of information required to compile distribution ranges is available for less than 25 per cent of species (60 per cent of Caribbean species, species of specific Indo-Pacific genera, and some of the common species of the other genera). Figure 3.4.1. displays an alternative that may contain errors within regions, but is unlikely to contain significant error at the level of inter-regional comparisons. Unfortunately, it is not possible to generate meaningful regional classifications with such data on Figure 3.6.7.

3.4.2. Species

Within the tropical Indian Ocean, species diversity is so uniform that Figure 3.4.1. has a pattern of contours similar to those generated from family-level and generic-level data (see Figure 3.6.7.), respectively.

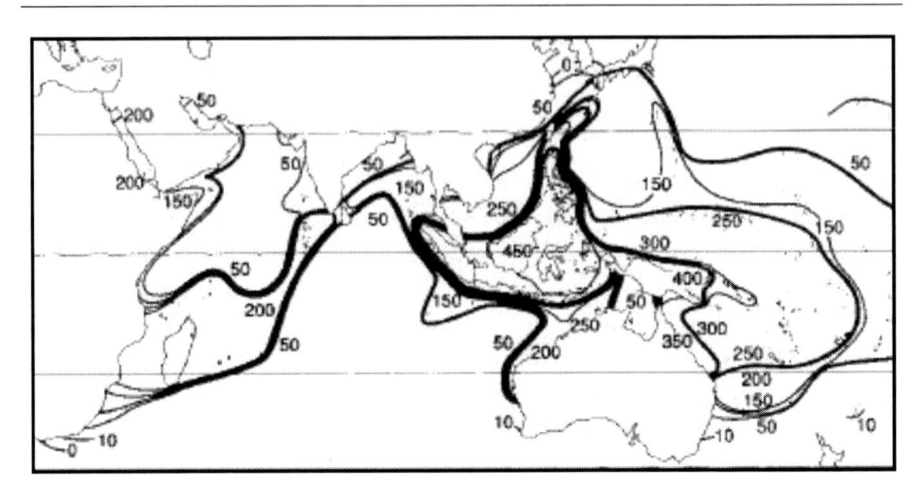

Figure 3.4.1. The distribution ranges of all genera by number of species in each genus (by Veron, 1995).

Finer contouring (not shown) indicates slightly higher species diversity in the central Red Sea than elsewhere in the province.

The boundaries of the Indo-Pacific center of diversity are mostly well defined: Sumatra and Java in the southwest; Sabah and the Philippines in the northwest; the Philippines, eastern Indonesia and Papua New Guinea in the northeast.

Species diversity of the Great Barrier Reef is intermediate between that of the Indonesia -Philippines Archipelagos and the western tropical Pacific. East of the Great Barrier Reef, the tropical Pacific is sharply divided, by the 200 and 250 species contours, into a western province (the eastern part of the Indo-west Pacific province of Figure 49), and an eastern province.

A central and southeast Pacific province contains the 150 and 50 species contours. Other Pacific provinces, including Hawaii and the far eastern Pacific, are very peripheral at species level.

3.5. Distribution of Coral in Vietnam

According to the studies performed in the first decades of the twenty-first century, Vietnam's reef-building coral fauna comprised 376 species, pertaining to 80 genera (including 9 ahermatypic corals), of which 137 species, belonging to 26 genera, were not previously known for that region, and 13 species from 6 genera were described for the first time). As in most

Indo-Pacific reefs, the species diversity of Vietnam's reefs consists mainly of the members of 5 families, Acroporidae (117 species), Faviidae (42 species), Fungiidae (32 species), Poritidae (31 species), and Dendrophylliidae (25 species), making up 64.48% of the total *scleractinian* species composition (Figure 3.5.1). The five genera most diverse and widespread in all reefs comprise Acropora (90 species), Montipora (28 species), Porites (20 species), Favia (14 species), and Fungia (12 species) are most various and numerous on all reefs, making 47% of all specific riches of scleractinian (Figure 3.5.2).

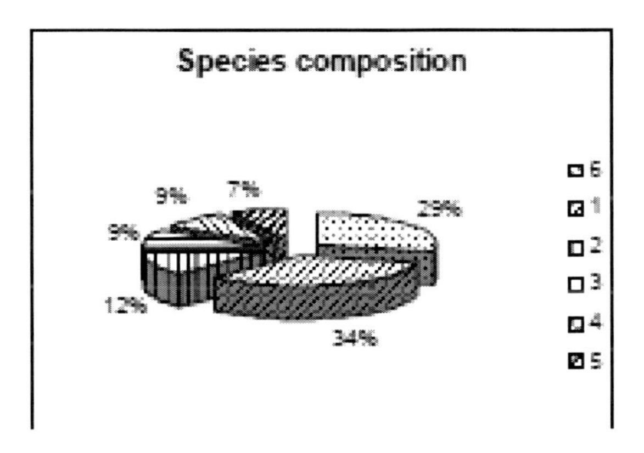

Figure 3.5.1. Relative species diversity of major coral genera in Vietnam. 1: Acroporidae; 2: Faviidae; 3: Fungiidae; 4: Poritidae; 5: Dendrophylliidae; 6: The rest genera.

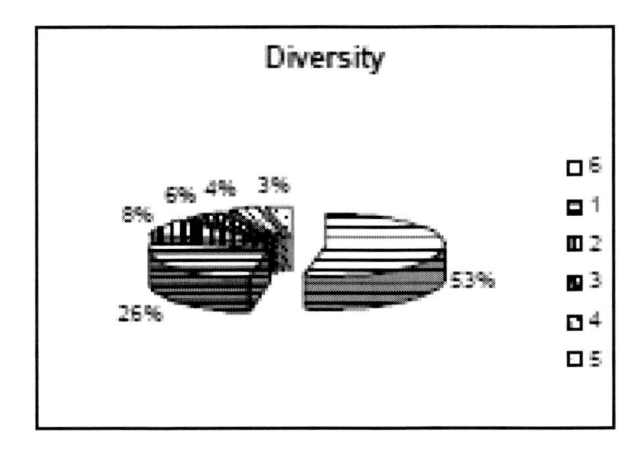

Figure 3.5.2. Relative species diversity of major coral species in Vietnam. 1: Acropora; 2: Montipora; 3: Porites; 4: Favia; 5: Fungia; 6: The rest species.

In all, some 20 scleractinian species form mono-specific settlements, varying from small "spots" (tens of square meters) to extended zones (hundreds of square meters), with coverage reaching 60%-100%. One fifth of all scleractinians occur throughout the Vietnam coast. As a whole, the species diversity of reef-building scleractinians in different areas of the Vietnam coast is quite comparable, ranging from 190 species in the Gulf of Tonkin to 265 in the South Vietnam. Similar (193 - 256) numbers of species were reported for reefs of Indonesia, the Philippines, and Western Australia [60]. Central and South Vietnam reefs are most similar in species composition and are quite comparable to Spratly reefs.

Both gulfs are shallows with high water eutrophication and turbidity, with a predominance of clay sediments. These factors cause a similarity of the biological and morphostructural zonation of reefs and species composition of reef communities in the gulfs. At the same time, certain differences in climatic and geomorphological conditions of the gulfs result in some dissimilarities in their scleractinian species composition. The development, zonation, species composition, and structure of the reefs in the gulfs were reported previously [49, 22, 25], so here, only major similarities and differences will be mentioned. The reef communities in both gulfs lack members of *Palauastrea* and *Caulastrea* and *Acropora cuneata*, the latter occurring in most Vietnam reefs. *Plerogyra* and *Physogyra* are absent in the closed part of the Gulf of Tonkin, and *Pachyseris, Mycedium*, and *Pectinia*, in the innermost and coastal areas of the Gulf of Thailand. However, some members of the latter three genera and rare *Physogyra* and *Plerogyra* species are found in the open parts of both gulfs, of Hainan and Tho Chu islands.

Corals having large polyp forms and capable of self cleaning - *Galaxea, Echinopora, Lobophyllia, Echinophyllia, Turbinaria, Podobacia, Lithophyllon, Fungia*, and *Goniopora* - are widespread in both gulfs. The reefs in both gulfs are dominated by many species of these genera (*Galaxea fascicularis, Goniopora stokesi, Echinopora lamellosa*, and *Lobophyllia hemprichii*), as well as by *Acropora cytherea, A. nobilis, Montipora hispida, Porites lobata*, and *P. cylindrica*, widespread in Indo-Pacific reefs. Altogether, these species cover 60%-80% of the substrate. Massive *Porites* colonies (at least 10 species) forming vast monospecific settlements are typical of both gulfs. At the same time, members of *Pocillopora*, abundant in most Indo-Pacific reefs (5-7 species), rarely occur in the innermost parts of the gulfs (2 species maximum) but are common for island reefs in the open parts of the gulfs (Tho Chu, Hainan). Largely, the two gulfs are quite similar in coral diversity and share 74.3% of species.

Chapter 4

REEF ECOSYSTEM OF VIETNAM

4.1. REEFS OF VIETNAM AS A PART OF THE PACIFIC REEF ECOSYSTEM

The coastline of Vietnam is over 3200 km long and covers 15 degrees of latitude, from the Gulf of Siam in the south (8°N) to the Chinese border in the north (23°N). The near-shore water area (up to 50-m deep) of Vietnam, including some 3000 islands, is about 206000 km^2. Vietnam and its coastline are divided into 5 parts, the Gulf of Tonkin, Central and Southern Vietnam, Gulf of Siam, and Spratly Islands [56]. Reef-building corals and reef accumulations are confined to hard grounds, typical of the Vietnam coast. Between 16° and 19° N, the coastline is formed mostly by moving sand with a minor presence of hard substrates. The temperature varies between 18–32°C, and the salinity, 28–40%. One hundred and fourteen rivers are registered along the coastline. The spread of the reef is limited near the mouths of two large rivers, the Red River in the north and the Mekong in the south, due to adverse conditions. The ecosystems of the coral reefs of Vietnam feature high bioproductivity, with a primary production of up to 30–100 mg C/m^3 per day, which is almost 100 times that in open waters [1].

Vietnam is situated in the tropics, affected by two sorts of monsoons: the wet southwest, lasting from May till September, and the dry northeast, occurring in October–April. Heavy rain showers during the wet monsoon period result in a huge (5–400 million m^3) freshwater influx and a substantial (up to 200 000 tons) terrigenous sediment influx into the sea. The daily-suspended matter precipitation rate in the reef reaches 70–100 g/m^2 and increases tenfold during typhoons. This results in a remarkable decrease in water transparency, affecting, together with other factors, the development of coral settlement in this region.

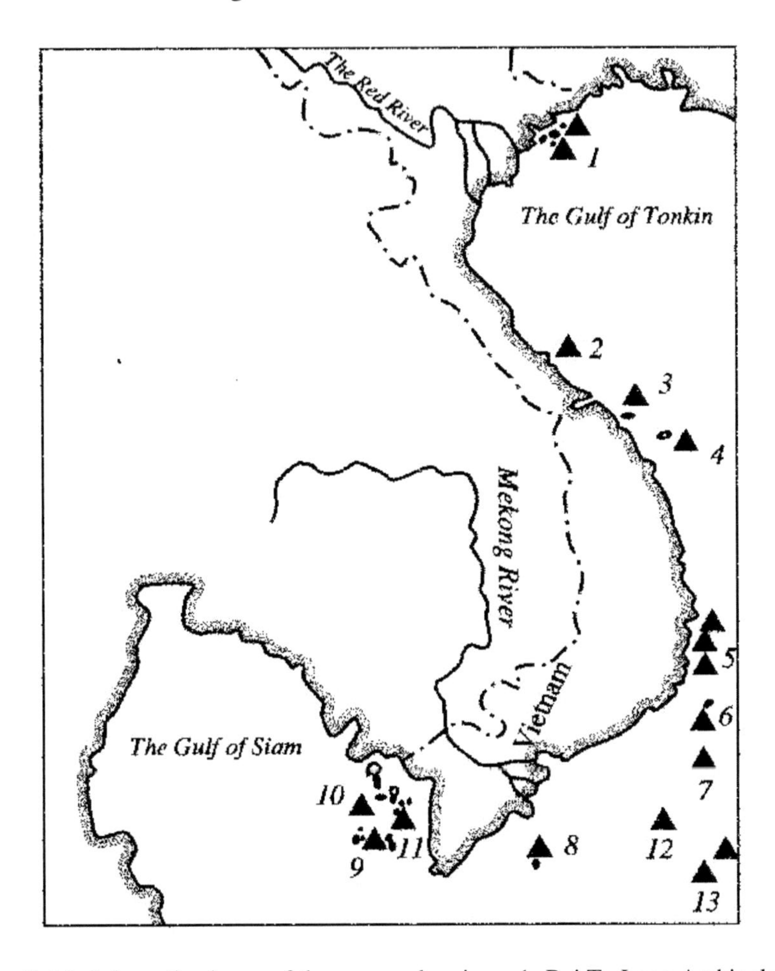

Figure 3.6.1. Schematized map of the surveyed regions. 1- Bai Tu Long Archipelago, 2 - Ze Island, 3 – Cape Danang, Cham and Son Tra islands, 4 – Ly Son Island, 5 - reefs of the Khanh Hoa Province, 6 - Thu Island, 7 - Ca Thuik Islands, 8 - Con Dao Islands, 9 - Tho Chu Island, 10 – An Thoi Archipelago and Namsu Islands, 11- Rach Gia bay, 12 -Royal Bishop and Astrolab shoals, 13 - Spratly Islands.

The reef-building corals and reefs of Vietnam attracted scientific attention as early as the first half of the twentieth century [51, 10, 39]. T. Loi Who? was the first to analyze the zonation of reef-building corals in reefs of the Khanh Hoa province [39]. He distinguished four scleractinian-dominated facies. These investigators determined the species composition of scleractinian and demonstrated its similarity to that of Australia and Indonesia. Beginning in 1980, systematic studies of Vietnam corals and reefs were performed in joint expeditions by the Institute of Marine Biology (Vladivostok), Nha Trang Institute of Oceanography, Haiphong Institute of Oceanology, and WWF (World Wide Fund for Nature). The published results were mainly related to scleractinian composition and distribution, with some papers analyzing the common accompanying macrobenthos species and a few publications providing the general characteristics of the reefs. Part of the data obtained was presented only in unpublished reports. Some findings were published in difficult-to-obtain regional works, including Vietnamese ones.

To date, in a region bordered by the Gulf of Tonkin in the north, the Gulf of Siam in the south, and the Spratly Islands in the South China Sea, all reef-building areas including large islands and shoals have been studied (Figure 3.6.1).

It is thus topical to review the data available on the coral composition, distribution, and development of Vietnam's reefs, as well as on their relation to the Indo-Pacific reef ecosystem. The rapid growth of the country's population of 70 million, together with travel industry development and marine aquaculture intensification, has resulted in increased anthropogenic pressure on this unique ecosystem. Generalizing data on the composition, structure, and environmental conditions of Vietnam's reef communities will allow one to evaluate the degree of their degradation and the outlook for their conservation and recovery.

Both along the coastline and around the islands, reef-building corals form diverse reef topographies. These include small fringing reefs along the coastline, barrier reefs separated from the continent (Ly Son Island and Giang Bo Reef), and atolls (Spratly Islands) in the open South China Sea [21, 65, 34]. Reported Vietnam's reefs include both true reef frameworks and coral gardens. Various calcareous structures occurring on reefs might be formed by coral settlements, usually called coral layers, communities, or specialized settlements. Such structures are typical of early reef development stages and lack any geomorphological and vertical biological zonation. All Vietnam reefs display distinct biological and more or less prominent morphostructural zonation. By the specificity of geomorphological and climate conditions, Vietnam's fringing reefs are clearly distinguished into two types (Figure 3.6.2).

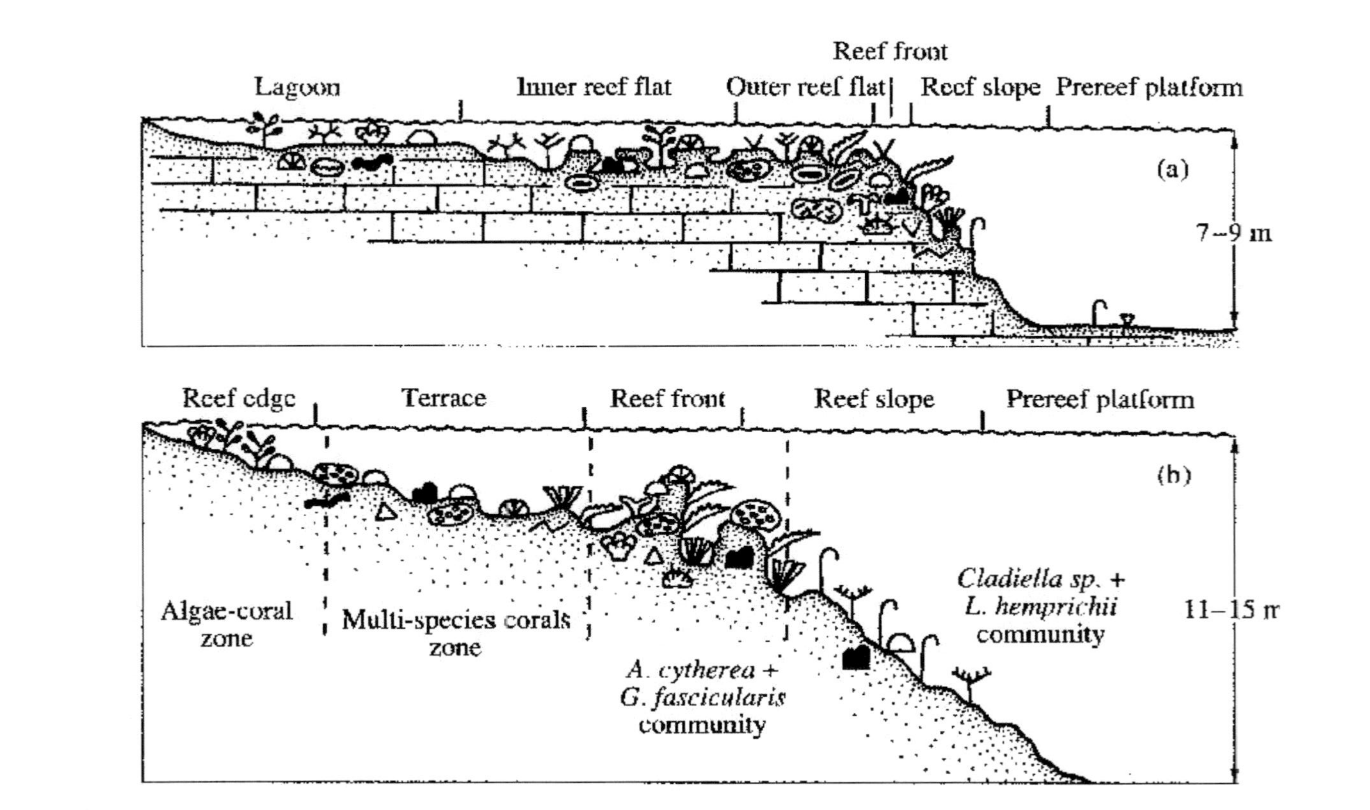

Figure 3.6.2. Schematized profiles of structural (a) and unstructured (b) reefs. See the text for comments.

Reefs of the first type have a distinct zonation (reef lagoon, reef-flat, etc.) with a developed carbonate framework, so-called structural reefs common for the tropical zone of the World Ocean. Reefs of the second type display a weak morphostructural zonation, with some zones occasionally absent. Carbonate deposits in such reefs comprise only coral settlements of a low thickness, hardly changing the overall substrate profile. These are so-called structureless or encrusting (Latypov, 1995) reefs. Vietnam's structural reefs are mainly formed in closed bights and on the organogenic base of Holocene reefs, while structureless reefs are formed off promontories and in open bights, mostly on stone and rocky substrates [23, 65, 66].

Vietnam's reefs pertain to the epicontinental monsoon type. They are situated at the South China Sea periphery. The shoal waters of this region are highly eutrophicated, and the grounds are highly silted due to the huge amount of terrigenous influx. Other hydrological conditions are also not optimal for reef-building coral growth. Thus, in the Gulf of Tonkin, the salinity may drop to 26% and winter water temperature to 16°C. Heavy northeast monsoon winds generate coastal waves up to three meters high with a 6-s period. During southwest monsoons, the Vietnam coast is struck by 5–10 typhoons per year. Vietnam's reefs feature a moderate vertical and horizontal spread and low thickness of modern reef-derived deposits. Their offshore spread usually does not exceed 200–300 m. They rarely extend to a depth of over 20 m. Sometimes they lack distinct morphological zonation. Most of Vietnam's reefs have an indistinct reef flat and slope. In some reefs, mostly ones on stone and boulder substrates, the only distinct zone is the reef slope. However, they all have a distinct vertical biological zonation, showing up in the dominant species succession and in the change in the composition and structure of coral communities and accompanying macrobenthos.

According to the studies performed in the last decades of the twentieth century, Vietnam's reef-building coral fauna comprises 382 species, pertaining to 80 genera (including 9 ahermatypic corals), of which 131 species, belonging to 26 genera, were not previously known for that region, and 12 species from 6 genera were described for the first time [21, 33, 63]. As in most Indo-Pacific species, the diversity of Vietnam's reefs consist mainly of the members of 5 families, Acroporidae (110 species), Faviidae (38 species), Fungiidae (32 species), Poritidae (31 species), and Dendrophylliidae (25 species), making up altogether 64.48% of the total scleractinian species composition.

The five genera most diverse and widespread in all reefs comprise *Acropora* (56 species), *Montipora* (28 species), *Porites* (18 species), *Favia* (13 species), and *Fungia* (11 species). In all, some 20 scleractinian species form monospecific settlements, varying from small "spots" (tens of square meters) to extended zones (hundreds of square meters), with a coverage reaching 60–100%. One fifth of all scleractinian occur throughout the Vietnam coast. As a whole, the species diversity of reef-building scleractinian in different areas of the Vietnam coast is quite comparable, ranging from 190 species in the Gulf of Tonkin to 265 in South Vietnam (Figure 3.6.4). Similar (193–256) numbers of species were reported for reefs of Indonesia, the Philippines, and Western Australia. Central and South Vietnam reefs are similar in species composition and are quite comparable to Spratly reefs.

The peculiarity of the coral faunas of the Thailand and Tonkin gulfs, as revealed by cluster analysis (Figure 3.6.5), is consistent with their ecological peculiarities. Their scleractinian diversity is partly caused by their similar hydrological regimes. Both gulfs are shallows with high water eutrophication and turbidity, with a predominance of clay sediments.

These factors cause a similarity of the biological and morphostructural zonation of reefs and species composition of reef communities in the gulfs. At the same time, certain differences in climatic and geomorphological conditions of the gulfs result in some dissimilarities in their scleractinian species composition. The development, zonation, species composition, and structure of the reefs in the gulfs were reported previously, so only major similarities and differences will be mentioned here.

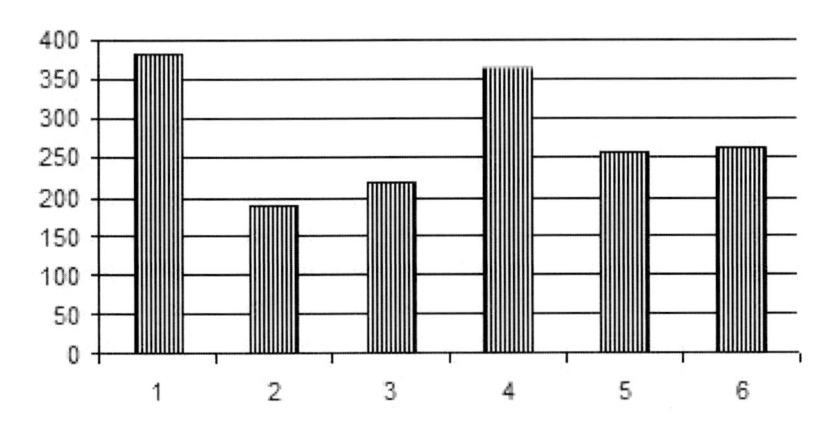

Figure 3.6.4. Scleractinian species diversity in different regions of Vietnam. 1-total number of species (382), 2-Gulf of Tonkin (190), 3-Central Vietnam (219), 4-South Vietnam (365), 5-Gulf of Thailand (256), 6-Spratly Island (261).

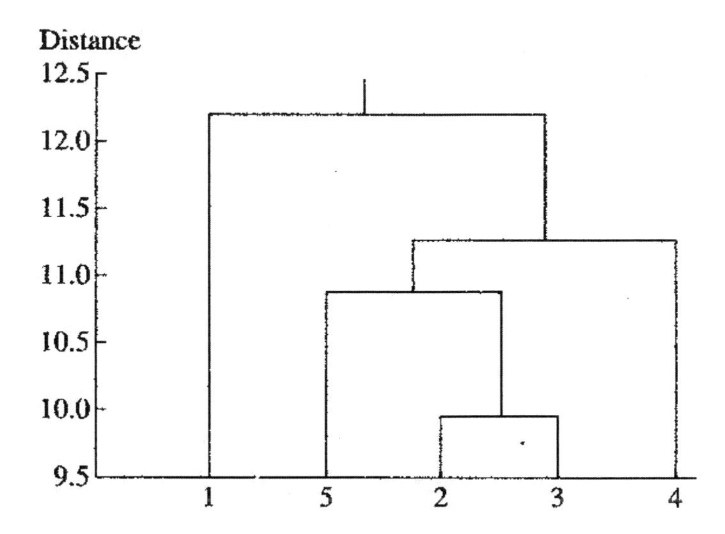

Figure 3.6.5. Similarity dendrogram of scleractinian faunas in different regions of Vietnam. 1 - Gulf of Tonkin, 2 - Central Vietnam, 3 - South Vietnam, 4 - Gulf of Thailand, 5 - Spratly Island.

The reef communities in both gulfs lack members of *Palauastrea* and *Caulastrea* and *Acropora cuneata*, the latter occurring in most Vietnam reefs. *Plerogyra* and *Physogyra* are absent in the closed part of the Gulf of Tonkin, and *Pachyseris, Mycedium*, and *Pectinia*, in the innermost and coastal areas of the Gulf of Thailand. However, some members of the latter three genera and rare *Physogyra* and *Plerogyra* species are found in the open parts of both gulfs, off Hainan and Tho Chu islands. Corals having large polyp forms and are capable of self-cleaning - *Galaxea, Echinopora, Lobophyllia, Echinophyllia, Turbinaria, Podobacia, Lithophyllon, Fungia*, and *Goniopora* - are widespread in both gulfs. The reefs in both gulfs are dominated by many species of these genera (*Galaxea fascicularis, Goniopora stokesi, Echinopora lamellosa*, and *Lobophyllia hemprichii*), as well as by *Acropora cytherea, A. nobilis, Montipora hispida, Porites lobata*, and *P. cylindrica*, widespread in Indo-Pacific reefs. Altogether, these species cover 60–80% of the substrate. Massive *Porites* colonies (at least 10 species) forming vast monospecific settlements are typical of both gulfs. At the same time, members of *Pocillopora*, abundant in most Indo-Pacific reefs (5-7 species), only rarely occur in the innermost parts of the gulfs (2 species maximum) but are common for island reefs in the open parts of the gulfs (Tho Chu, Hainan). The two gulfs are quite similar in coral diversity and share 74.3% of species.

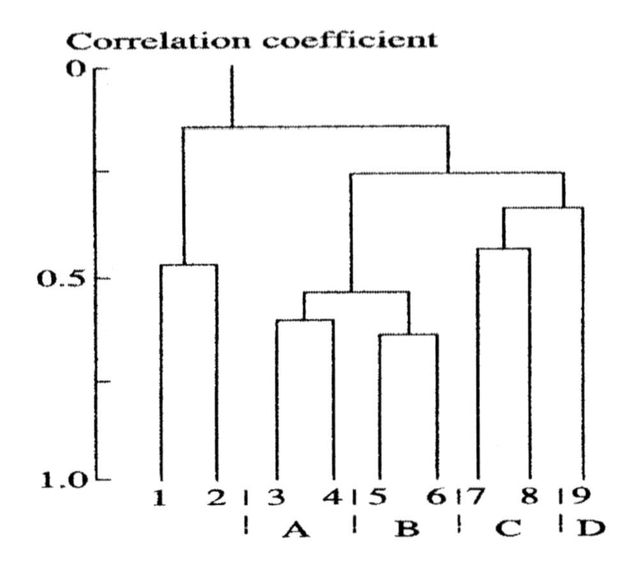

Figure 3.6.6. Cluster dendrogram of the species composition of benthic communities. Algal–coral community, unstructured (1) and structural (2) reefs; Acropora community, unstructured (3) and structural (4) reefs; Acropora+Diploastrea community, unstructured (5) and structural (6) reefs; (7) bioherm community (reef slope); (8) Junceella+Diaseris community; (9) Maleus+Junceella community. A - lagoon, B - reef flat and terrace zone, C - reef slope, D - for-reef platform.

The distribution and peculiarities of benthic communities in the coastal part of Vietnam reefs is rather constant (see Figure 3.6.2). As a rule, these are algal–coral communities, composed of several biocoenoses (zones, facies), and dominated by individual algal or coral species or by groups of species. The predominance of *Laurencia, Turbinaria*, and *Sargassum* algae in the coastal zones of the reefs has been reported for many reef development areas. This may be indicative of an increase in water eutrophication or later stages of reef development. Both along the Vietnam coast and in the whole Indo-Pacific, in reef zones characterized by relatively stable conditions (lagoons, deep stony and coral terraces, and reef slopes), branched, plate, and trumpet colonies of *A. cytherea, A. hyacinthus, Montipora danae, M. foliosa, Porites cylindrica, P. nigrescens*, and others successfully compete with differently shaped scleractinian colonies. A wider distribution of encrusting and plate colonies of *Euphyllia, Echinophyllia, Mycedium, Pachyseris*, and *Turbinaria* compared to that of branched forms is directly caused by lowered illumination. This is also the case in many reefs of the Indo-Pacific and Caribbean. In Vietnam's reefs, such corals are common for communities of the slope base, bioherms zone, and for reef platform.

Caused by *abiotic* factors, the vertical distribution of reef-building corals has a strong effect on the development of biotic zonation across a whole reef community, beginning with settlement-site choice and ending with interspecific trophic relationships. The relationships between the species composition of benthic communities of some reefs as revealed by cluster analysis correlated with the ecological and physiographical zonation of the reefs (Figure 3.6.6).

Algal–coral lagoon and reef flat communities dominated by red and brown algae and similar in coral and common macrobenthos species composition formed a single cluster group, that of communities developing under similar conditions. The high similarity between coral faunas from different sites reflected similar, sometimes extreme, conditions of reef flat and shallow-water stone terrace. At the same time, communities of these reef zones were set apart from those of neighboring reef zones. Both in structural and unstructured reefs, polyspecific reef slope communities sharing a relatively greater number of corals also formed a distinct cluster.

To summarize the above, both structural and unstructured reefs feature vertical biological and geomorphological zonations. The latter is mainly determined by peculiarities of the underwater reef slope substrate. Similar biological zonation reflecting interzonal differences in environmental conditions (substrate, wave regime, sedimentation rate, illumination) has been reported for many of Vietnam's reefs and various Pacific and Caribbean reefs.

Shallow-water Vietnam reefs growing in highly eutrophic conditions lack thick reef deposits and feature high coral diversity and distinct biological zonation, that is, the presence of inner heterotrophic (lagoon, reef flat) and outer autotrophic (reef slope) zones, which is characteristic of typical Indo-Pacific reefs. In reefs of Indonesia and Philippines and in the Great Barrier Reef, a total of 360–410 reef-building scleractinian pertaining to 70 genera have been recorded. This region of the Western Pacific is considered the center of origin of tropical coral faunas. The maximum coral diversity is observed in the so-called Coral Triangle [12, 60] with apices in the Philippines, the Malacca Peninsula, and New Guinea (Figure 3.6.7).

Vietnam's reefs, too, obviously belong to this center, which is evidenced by their high similarity in coral species composition to reefs of Thailand, Indonesia, and the Philippines (76.4, 72.3, and 81.6%, respectively). In the greater Western Pacific Coral Triangle (with apices in Vietnam, South Japan, and the Great Barrier Reef), coral faunas are also highly similar and homogenous.

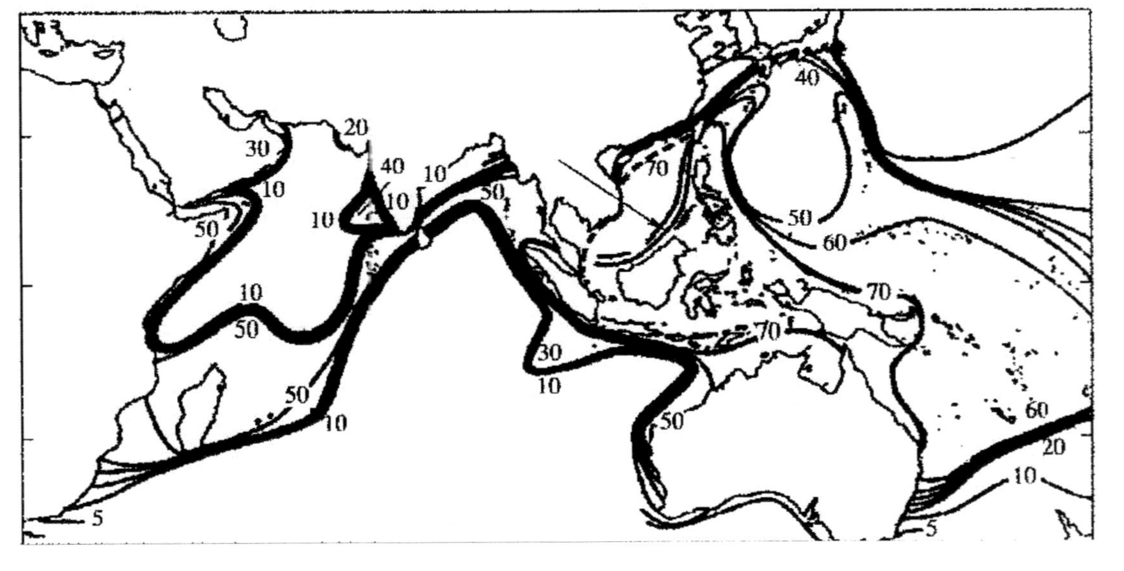

Figure 3.6.7. Schematized map of the generic diversity of of reef- building corals in different regions of the Indo-Pacific (partly after Veron, 1995). The dotted line and arrow indicate new and old 70 genera diversity isolines, respectively.

The similarities between the Vietnam coral fauna and those of Japan and Australia are 77.5 and 86%, respectively, suggesting homogeneity of the coral fauna of the Western and Southwest Pacific. As a whole, the species complex of Vietnam scleractinian, as well as those of alcyonarian and Gorgonarian, belongs to the tropical fauna, as the majority of Vietnam corals are also common for the equatorial Indo-Pacific reef zone. The scleractinian species composition of this area exceeds 80% of that of the Pacific, and the alcyonarian diversity of Vietnam's reefs is one of the greatest in the Indo-Pacific [23, 41].

The species composition and high diversity of Vietnam's coral fauna, as well as its close similarity to the Southwest Pacific coral fauna, allow one to refer it to the Indonesia–Polynesian center of origin of the coral faunas of the tropical Indo-Pacific. The whole Vietnam coast, from the Gulf of Tonkin to the Gulf of Siam, is a biogeographically single whole and is part of the Indo-Polynesian Province of the Indo-Pacific Area.

4.2. Coral Species and Structure of Vietnamese Reefs

4.2.1. Gulf of Tonkin

In the Gulf of Tonkin, two reef types occur—structural reefs with a distinct zonality and unstructured ones, too, with a distinct vertical zonality, appearing as the replacement of communities corresponding to those of structural reefs. In unstructured reefs, corallogenic deposits form a thin crust over the substrate, which hardly, if at all, changes its profile. The composition and distribution of the corresponding reef communities enable two reef zones to be distinguished, the inner heterotrophic and outer autotrophic ones. The zones form a single ecosystem and closely resemble those of reefs with a distinct zonality. For the above-described reefs, the term "encrusting reefs" was proposed as distinct from structural or developing reefs. Both encrusting and regular reefs occur all along the Vietnam coast, as well as other regions of reef formation. Their development is determined by the monsoon climate, which is characterized by continuous silting and periodic desalination, resulting in the perishing of many coral settlements. On the whole, the favorable period lasts somewhat longer than the unfavorable one enabling constant recovery of the encrusting reefs. In the middle-late 1990s, the data available on the reefs of the Gulf of Tonkin were considerably extended due to joint studies by Vietnamese and Russian scientists and by researchers of the World Wildlife Foundation (WWF). In all, about 50 reefs were studied in the northern, central, and eastern parts of the gulf (Figure 3.7.1).

The number of reef-building coral species was extended to 188 (see Table). Several new coral species were discovered. To date, the state of the art in the study of reefs of the Gulf of Tonkin is quite comparable with that of the Australian, Indonesian, and Philippine reefs. The studies resulted in a detailed description of the structure and composition of communities of structural and unstructured reefs. In each zone, dominant and subdominant coral and accompanying mass macrobenthos species were distinguished. In the near-shore zone of all reefs, an algal-coral community was found dominated by the red-brown sargassos, *Padina*, and *Turbinaria* algae. Rare corals formed small separate colonies in this zone. The reef flat zone of the structural reefs and the corresponding zone of the unstructured reefs were characterized by monospecific settlements of *Acropora* and *Montipora* —major reef-building corals of the whole tropical zone of the World Ocean. In the reef flat, the composition of corals and the structure of coral communities were nearly the same as those of many Indo-Pacific reefs [36, 66]. The peculiar nature of the reefs of the Gulf of Tonkin is determined by the monsoon climate in the region and, in the wintertime, the effects of runoff waters carried into the gulf by large and small rivers. These waters are cooled to 16-18°C, silted to 100 g/m^2 per day, and freshened to 28%. The reefs in the Gulf of Tonkin are thus exposed to conditions far from optimum for reef formation. However, the continuously arriving suspended matter does not settle out directly onto the coral settlements because of the huge integral water exchange and intense roiling. Water silting and eutrophication resulted in changes in the structure and composition of reef communities via the reduction or elimination of certain coral species. As a result, instead of acroporids, typical for the majority of other reefs, the communities of the reefs of the Gulf of Tonkin are dominated by Poritidae and Faviidae, which form the framework of the reefs. These peculiarities make the reefs of the Gulf of Tonkin really unique. The abundance of Poritidae was accounted for by their ability to secrete a firm mucous covering and start reproduction 1-2 months earlier than other coral species. These peculiarities favor their better adaptation to water eutrophication, overheating, and desalination under stressful conditions of silted shallow water. Massive colonies of Poritidae and Faviidae - one of the major bioproducents under the local conditions - not only form the reef framework but also play a considerable role in the expansion of the reef area.

Both biotic and abiotic factors cause the erosion of coral colonies, resulting in the passive colonization of vacant bottom areas by colony fragments. In the base zone of reef slope, a new, now organogenesis, substrate is formed, which is subsequently inhabited by both corals and other phyto- and zoobenthos species. These factors are important for the Gulf of Tonkin, a shallow-water gulf with the predominance of soft ground and a limited availability of solid substrate off shore. As compared to other reefs of Vietnam and many reefs of the Indo-Pacific, the reefs of the Gulf of Tonkin are characterized by a peculiar composition and structure of the reef communities.

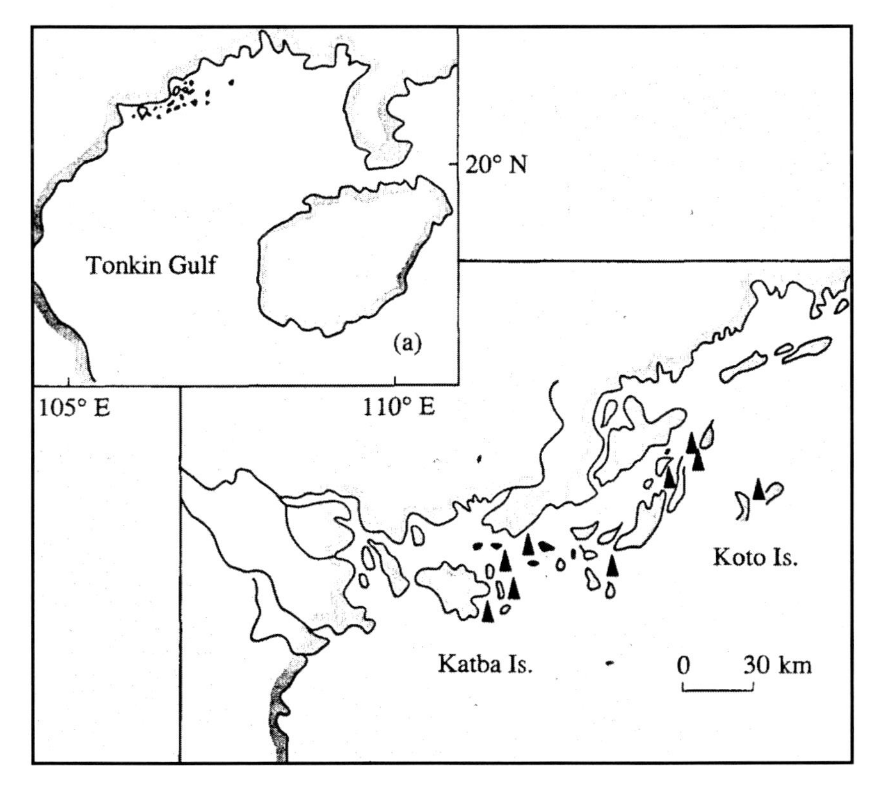

Figure 3.7.1. Schematized map of the Gulf of Tonkin (a) and the areas surveyed triangles.

These reefs develop under complex hydrobiological conditions. In addition, they are situated in a populous area and constitute constant food and, as a result of increasing tourism, financial sources for the local people. Thus, they may be and must be used as a model in the research on the conservation and recovery of reefs subject to stressful conditions and anthropogenic contamination. The peculiarities of the hydrological conditions of the Gulf of Tonkin and Gulf of Thailand cause the coral species composition. On one hand, the shallowness of both gulfs and high eutrophication and turbidity of their waters, caused by mainly clayey fractions, result in the similarity of their reefs and of the compositions of the reef communities to one another. On the other hand, the geographical remoteness of the two regions and the difference in their geomorphological conditions cause some differences in the composition of the coral communities of the gulfs. To date, members of the genera *Palauastrea* and *Caulastrea,* as well as *Acropora palifera,* common to most reefs, have been found in neither gulf. Members of the genera *Plerogyra* and *Physogyra* were not encountered in the Gulf of Tonkin, and members of *Pachyseris, Micedium,* and *Pectinia* have not been registered thus far in the innermost and near-shore parts of the Gulf of Thailand. However, some species of the latter three genera, as well as rarely occurring members of *Physogyra* and *Plerogyra* were found in the open parts of both gulfs, of Hai Nang and Tho Chu islands. It is noteworthy that the members of the genera distinguished for large polyp size - *Galaxea, Echinopora, Lobophyllia, Echinophyllia, Turbinaria, Podobacia, Lithophyllon, Fungia,* and *Goniopora*—were widespread in both gulfs. Most reef communities are dominated by numerous species of these genera, in particular, *Galaxea fascicularis, Goniopora stokesi, Echinopora lamellosa,* and *Lobophyllia hemprichii,* as well as by *A. cytherea, A. nobilis, 1. hispida, P. lobata,* and *P. cylindrica,* which are widespread in all Indo-Pacific reefs. The former species occupy as great an area as 60-80% of the substrate surface. Another feature shared by the two gulfs is the wide distribution of massive *Porites* colonies, forming vast monospecific settlements and exhibiting a huge diversity (not less than 10 species). As opposed to *Porites,* 5-7 *Pocillopora* species forming mass settlements in the majority of the Indo-Pacific reefs are extremely rare in the Gulf of Tonkin and Gulf of Thailand. No more than two species are to be found in the gulfs; the only exception are the reefs off Tho Chu and Hai Nam islands, situated in the open parts of the gulfs, where *Pocillopora* species are common. On the whole, the species compositions of scleractinian of the two gulfs are quite comparable, both qualitatively and quantitatively. The gulfs share 71.7% of their total numbers of scleractinian species. The differences between the gulfs in coral species composition are apparently due to the inequality in the progress of their study.

Promoted in the conservation of reefs as an integral part of the natural complex, a component of the national wealth of Vietnam, and the property of humankind, the governing body of the National Center for Natural Science and

Technology of Vietnam was informed that special attention should be given to the reefs of Bo Hung and Cong Do islands. The conservation and recovery of the high *biodiversity* of reef communities in these regions should be considered a first priority task in the framework of creating reserves and conservation areas in the Gulf of Tonkin.

4.2.2. Barrier and Platform Reefs

4.2.2.1. Giang Bo Reef

The bottom area adjoining the reef base at a depth of 17 m was a flat platform with silt–sand sediments with debris of invertebrate skeletons. The depth gradually decreased by 1.5–2 m from the 40th to the 100th meter of the transect. Massive and sporadic encrusting colonies of scleractinian occurred. Colonies of *Heliopora* occurred most often, *Pachyseris, Mycedium, Pavona, Montipora, Podobacia, Goniopora, Alveopora, Lobophyllum, Pectinia*, and *Euphyllia* were rather common, *Galaxea, Herpolitha, Fungia, Polyphyllia*, and large colonies *Porites* were sometimes recorded, and *Acropora* and small branching *Porites* occurred sporadically. The coral covering of the substrate was 7–10%. The associated macro fauna presented by singular individuals of the bivalve mollusks *Atrina vexillum* and *Malleus malleus* and of the sea stars *Linckia laevigata* and *Culcita novaeguineae.* The latter two species also recorded in the same numbers in all other zones of the reef. At the distance of 100–160 m from the beginning of the transect, the percentage of coral covering of substrate pronouncedly increased, from 10 up to 72%, with the same thing occurring to the size of colonies and species diversity of scleractinian *Acropora* most of all. The buttress system began from the 160th of the transect meter. The species diversity of the corals sharply increased (by 1.5 times) concurrently with the variety of their growth forms. The coral covering of substrate increased to 100%. Young settlements of scleractinian occurred rather often on an old lamellar colony of *Acropora* of a 100x60 cm area.

This zone extended almost to 70 m. The following approximately 40 m of the transect were a zone of *Acropora*, in which monospecies settlements of *A. cytherea*, *A. hyacinthus*, *A. formosa,* and *Porites rus* with individual colonies of *Goniastrea, Astreopora, Pocillopora, Montipora, Heliopora*, and massive forms of *Porites*, were clearly distinguished. At a depth of 5–3 m, at the 240–250th meter of transect, the reef flat zone extending by 200–300 m began with individual channels. It characterized by a continuous covering of substrate by corals (Figure 3.7.2) and interrupted by hollows oriented from northeast to southwest.

The ample tangled settlements, up to 100–150 m wide, were composed of *A. cytherea, A. hyacinthus,A. formosa, P. rus, Montipora aequituberculata*, the seaweed *Chnoospora implexa* and sea mat *Zoanthus* sp. Individual colonies of scleractinian and small bioherms spread in channels of sand sediments. The holothurian *Holothuria atra* (0.2 ind./m^2), the bivalves *Tridacna crocea* (up to 2 ind./m^2), and *T. squamosa* (up to 0.1 ind./m^2) were the most abundant of the associated fauna, including the *gastropods Cymatium* spp. and *Oliva* spp. (up to 3 ind./m^2) occurring in the sand. Corals were associated with bivalves, with domination of *Arca ventricosa*(10−15 ind./m^2) and *Beguina semiorbiculata* (up to 10 ind./m^2), common *Septifer bilocularis, Pinctada margaritifera* and *Cardita variegata*, as well as the gastropods *Tectus niloticus, Trochus maculates* and *Turbo petholatus*. The projective cover of substrate by corals varied from 10 up to 100% and included more than 100 species in its composition.

Figure 3.7.2. Monospecies population of Acropa on reef flat of the reef Giang Bo (depth 1 m).

The west–east transect. At the beginning of the transect, oriented from the shore to the open sea, at depth of 4.5 m, an extended scrap area of dead corals was observed. Small, up to 10 cm diameter, individual colonies of *Pocillopora, Stylophora, Acropora (A. palifera, A. humilis, A. millepora,* and *A. florida), Montipora, Porites, Leptoria, Platygyra, Pavona, Hydnopohora, Favia*, and *Favites* occurred for the next 50 m. The projective coral covering the bottom was 10–15%. There were also numerous dead lamellar colonies of *Acropora* up to 2 m in size. Between the 140th and 150th meter of the transect *Stylophora pistillata* dominated among corals in its frequency of occurrence. Further, the zone of settlements of *Acropora, Porites* and the alga *Chnoospora* was spread; the projective cover of the bottom was about 75%. The scleractinian habitus showed that they were under attack by the sea star *Acantaster planci,* whose population density reached 5–7 ind/m^2.

Concurrently with *A. planci* the gastropod *Charonia tritonis*, the bivalves *T. squamosa*, and relatively young *T. crocea* occasionally occurred there. At a distance of 200–400 m from the beginning of the transect the settlements of *Acropora* were still spread, however, the settlements of *A. formosa, A. cytherea*, and *A. hyacinthus* were the most common there. Among *Acropora* settlements, more often than previously, bottom areas with dead corals and their fragments, which were more or less covered with the calcareous algae *Amphizoa, Halimeda*, and *Hyphnea panosa* occurred. With the 430th meter of the transect *Acropora* settlements acquired a distinctly patched pattern, and the percentage of projective cover of the bottom varied from 40 up to 80%. In that zone as a whole, concurrently with *Acropora*, individual colonies of *Favia, Favites, Montipora, Porites* and *Heliopora* were widespread. Starting from the 450th meter of transect, the extended buttress system area began at a depth of 10 m, comparable to that of the south reef area. It specified by the same set of growth forms and a similar species composition of corals and associated fauna. The zone was traced along the transect up to the 520th meter, then the number of corals and diversity of their growth forms decreased on the reef slope. Massive and encrusting forms dominated and occasional small bioherms occurred. The projective cover of the bottom did not exceed 12–15%. A further increase in depth up to 14 m was associated with strong silting of the bottom and an increase in water turbidity. Only occasionally, after the 570th meter of transect, was it possible feel a small colony of coral by hand.

4.2.2.2. Reef of Bach Long Vi Island

Bach Long Vi Island extends for 1.6 km from northeast to southwest at the exit from the Gulf of Tonkin. It is surrounded by an adjoining reef with well-defined main physiographic zones (reef crest, reef flat, reef slope, etc.) and a small lagoon on its southwestern side. In common, (264) species of corals and accompanying macrobenthos were identified on the investigated reef. Representatives of the Acroporidae, Poritidae, and Mussidae predominated among scleractinian. Single-species aggregations of alcyonarian *Sinularia* and *Lobophyton* and the hydroid *Millepora* were fairly numerous. Of the accompanying macrobenthos, the sea urchins *Echinostrephus molaris, Diadema setosum* and *Echinothrix diadema*, the starfishes *Acantaster planci* and *Culcita novaeguineae*, as well as the holothurian *Holothuria atra* occurred most frequently. The bivalves *Lopha cristagalli* and *Pinctada margaritifera* were almost ubiquitous. *Asparagopsis taxiformis, Caulerpa racemosa, Turbinaria ornata, Padina australis, Sargassum olygocystum*, and *Gracilaria* sp. predominated among the macrophytes.

At the beginning of transect, on the northeastern part of the reef, at a depth of 16 m is a fore reef platform or a fore reef. This is an area of slightly silty coarse and medium-grained sand with inclusions of blocks and fragments of dead coral. Sparse colonies of alcyonarian, gorgonians, and scleractinian were encountered here. A gentle slope was observed stretching 250–300 m northwest. With decreasing depth, the number of large colonies of corals markedly increased. Soft corals *Sinularia dura, Lobophytum* sp., *Sarcophyton* sp., scleractinian *Favites abdita, Isopora palifera, Montipora venosa, Porites lobata*, and *Goniopora stokesi* predominated in terms of frequency. The coral covering of substrate was 5–7%. The reef slope down to about 12 m depth was covered with coral thickets and large bioherms up to 4–5 m high (Figure 3.7.3).

This reef zone was dominated by two to three species of *Sinularia, Lobophytum*, and *Sarcophyton* (Figure 3.7.4), massive colonies of *Porites*, plate-like and encrusting *Montipora, Pectinia, Pachyseris*, plate-like and digitate *Acropora*, and branching *Pocillopora*. The macrophytes *Turbinaria, Asparagopsis* and *Gracilaria* were also encountered.

The covering of substrate was up to 30–40% for corals and 10–20% for algae. Large colonies of *Porites* provided the basis for bioherms. Among other common inhabitants of the reef were the starfishes *A. planci, C. novaeguineae*, and *L. laevigata*, the sea urchin *D. setosum*, and mollusks *P. margaritifera, Pteria penguin, Tridacna squamosa, Lambis lambis* and *Conus textile.* With decreasing depth to 10–8 m, the number of corals increased and hence so did the coral cover up to 7-10%. The most widespread on this part of the reef slope were *S. dura, Acropora humilis, M. venosa, P. lobata, G. stokesi, Platygyra lamellina, Leptoria phrygya, Lobophyllia hemprichii, Merulina ampliata* and the hydroid coral *Millepora platyphylla* (coral cover 40–60%). Individual bushes of macrophytes *P. australis, Turbinaria* sp. and *A. taxiformis*, the sponge *P. testudinaria*, the starfish *L. laevigata*, the sea urchin *E. diadema*, and the holothurian *H. atra* were encountered.

Figure 3.7.3. Bioherm on the reef Bath Long Vi (depth 10 m).

Figure 3.7.4. Polyspecific settlement of various Alcyonarian on the reef Bath Long Vi (depth 10 m).

At a depth of 6–2 m, the upper reef slope with a well-defined system of spurs and grooves (buttress system) supports the highest species diversity of scleractinian (more than 150 species). A community of *Acropora* and *Montipora* occurred at this depth. *Acropora cytherea, A. hyacinthus, A. formosa, A. grandis, Montipora hispida, M. aequituberculata, M. vietnamensis, Pocillopora verrucosa* and *Seriatopora hystrix* were dominant. As a rule, large colonies of *Acropora millepora, A. specifera* and *M. hispida* (or *M. eaequituberculata*) formed stacked aggregations that covered 100% of the substrate over several tens of square meters. In this zone, the calcareous algae *Poroliton* sp. and *Halimeda* sp., the alcyonarian *Cladiella pachyclados*, the scleractinian *Acropora robusta, A. humilis, A. monticulosa, I. palifera, Pocillopora woodjonessi, S. hystrix* and *Galaxea astreata*, various poritids and faviids (5–7 species of each), the hydroid *M. platyphylla*, macrophytes *Turbinaria, Gracilaria* and *Asparagopsis* were widespread. Along with them, the starfishes *L. laevigata, C. novaeguineae*, the sea urchins *D. setosum, E. diadema*, the holothurians *H. atra,* and the individual mollusks *T. squamosa, Trochus maculatus, Cyprea arabica, and C. textile* were encountered.

The accompanying fauna on the northwestern part of the reef was comprised of the same set of species as in the corresponding zones on the northeastern reef side. However, a greater number of mollusks, holothurians, sea urchins, and other organisms were found here.

With regard to its geomorphological profiles, coral species diversity and zonal distribution, the reef of Giang Bo, Re Island and Bach Long Vi Island are comparable to the ribbon reefs of the Great Barrier Reef in Australia and to the barrier reefs of the Philippines and Indian Ocean [45, 46]. Based on its geomorphological characteristic, the presence of a large elongate and wide reef flat with distinctive flora and fauna, as well as a small lagoon, the investigated reef can be classified accordingly as the barrier reefs and the platform reef type.

The Bach Long Vi reef exhibits a pronounced axial zoning. Its central part (reef flat) was characterized by a homogenous coral community distinctly dominated by branching and plate-like *Acropora* and *Montipora*, which formed extensive fields. The periphery of the reef plateau consists of an extensive zone of accumulation of debris from the remains of dead coral, sand, and silt. The reef community is approaching a maturity or equilibrium stage analogous to similar Indo-Pacific reefs [8, 25].

The high species diversity of the reef community of Bach Long Vi and its ecological status requires special attention, and therefore recommendations were given to the Academy of Science and Technology of Vietnam regarding the establishment of a protected area on the Bach Long Vi reef for the purpose of conservation and restoration of the biodiversity of the Gulf of Tonkin as a whole.

4.2.3. Open part of the South China Sea

Survey results for most of the Vietnamese reefs were given in a number of publications of the one author of this book (Latypov, 2003a, b, 2011). In the last decade of the past century and in the first decade of the present century, Vietnamese reefs in the open part of the South China Sea were studied, particularly at the islands of Ca Thuik, Con Dao, Thu and Lodd in the Spratly Archipelago (see Figure 3.6.1). Below there is a generalized characteristic of coral communities of these reefs. Quick growth of the 70 million population of Vietnam, developing travel business, intensification of marine culture economies increase the anthropogenic pressure on this unique ecosystem even more. Generalized data on the species composition, structure and existence conditions of coral communities of Vietnam allow evaluating the degree of their preservation or degradation, and preservation and recovery opportunities. There are physiographic zones well distinguished on the reefs of the open part of the South China Sea, as well as on the other reefs of Vietnam. They are well known reef zones (lagoon, reef flat etc.) and comparable zones: algae-coral, poly- or monospecific coral settlements. Lagoons, reef flat changing into well-marked reef slope develop in the bays with sandy or corallogenic shores. Coastal algae-coral zones with the species composition and structure of communities are well comparable to the communities existing in lagoons form at the rocky, stony shores. Lagoons or algae-coral zones are 10 to 80m long and up to 2-4m deep. The width of reef flat or zones of poly- and monospecific coral settlements usually make 50-100m with the depth of 3 to 12 m. Reef slope with the upper and lower parts, and slope platform on the surveyed reefs are morphologic and by the species composition and structure of the community settling, it is the same as most of the Indo-Pacific reefs. The width of this reef zone makes 20-70 m, depth drop is 6 to 40 m. Macrobenthos species diversity at all reefs was rather high, and varies from 315 to 355 species (including 228 species of corals). In general, scleractinian species diversity index was 25.6 (Figure 3.7.5).

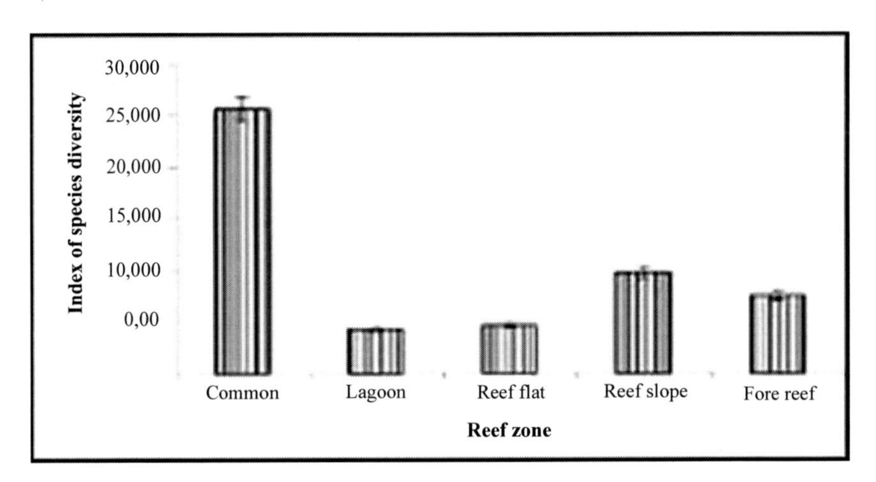

Figure 3.7.5. Index of species diversity in different zone reef.

The following species used to dominate there: Alcyonarian *Sarcophyton trocheliophorum, Lobophytum pauciflorum, Junceella fragilis scleractinian Seriatopora hystrix, Montipora grisea, M. hispida, M. aequituberculata, Isopora palifera, Acropora nobilis, A. cytherea, A. valenciennesi, Pachyseris rugosa, Porites cylindrica, P. rus, Goniopora stokesi,* mollusks *Arca ventricosa, Tridacna crocea, Malleus malleus, echinoderms Diadema setosum, Metalia sternalis, Brissus latecarinatus, Holothuria edulis, H. atra,* algae *Dictyota divaricata, Caulerpa racemosa, Sargassum polycystum, Padina australis, Laurencia obtusa.* Only three-five species of different macrobenthos representatives usually dominated at every reef. 41.2% of species composition of scleractinian has thorough distribution. Distinct lagoons formed only in rather big bays with sandy coasts and on the atolls of the Spratly Archipelago.

Sandy coast of lagoons are usually put to sea in a sloping manner, and some algae bushes *Ulva reticulata, P. australis, L. papilosa* and single colonies of Alcyonarian *S. trocheliophorum*, scleractinian *Seriatopora, Montipora, Acropora, Porites, Goniastrea* appear at the distance of 20-30m upon drawing away from the water cut. Algae may form thick brushes (up to 352 spec/m^2 and biomass 4700 g/m^2). At the distance of 50 minds farther, the amount of algae and corals usually grows. There are rare settlements of *S. hystrix* (on the area of up to 2x2m) and colonies of *I. palifera.* Macrophytes play a noticeable role in the structure of community: *Turbinaria ornata* (15-40% of biomass), C*aulerpa racemosa* (10-30%), *Sargassum duplicatum* (7-15%). In the benthos community of lagoons, the most numerous corals are *Acropora hyacinthus, A. florida, M. aequituberculata, M. hispida, Platygira* *daedalea, Leptoria phrygia, Porites rus, Pavona decussata* and *Hydnophora exesa*. There are rare hydroids *Millepora platyphylla, M. dichotoma*. The degree of substrate coverage with algae and corals reaches 40-60%. The following representatives of large macrobenthos make a constant component of algae-coral community of lagoons: mollusks. *Lambis lambis, Trochus niloticus, Cyprea arabica, C. tigris,* echinoderms *Linckia laevigata, D. setosum, H. edulis* and H. *atra*, which may have population density up to 10-15 spec/m^2. In the wide lagoons of the islands Thu, Con Dao and atoll Lodd there are mono settlements of colonies *A. valenciennesi, A. microphthalma* spread over dozens of square meters (Figure 3.7.6).

Figure 3.7.6. Monospecific settlement coral Acropora in the lagoon of the island Thu, depth 2 m.

Domination of two species of staghorn corals in this zone differs from other zones by a low index of species diversity 4.3. Remains of colonies and live corals, cemented chalky algae, organogenic detritus and fragments of branchy corals make up the reef flat. There is a developed system of longitudinal and transversal channels; the latter may reach 2-6m wide and 15-40m long. High level of coverage with corals (up to 50%) is ensured by monospecific settlements *of Acropora humilis, A. digitifera, A. microphthalma, A. nasuta, Porites cylindrica, M. foliosa,* hydroids *M. platyphylla, M. dichotoma,* chalky algae *Halimeda opuntia*. In channels, colonies and bioherms consisting of different species of *Acropora, Porites* (massive and branchy forms) *Seriatopora, Stylophora, Psammocora, Hydnophora, Montipora, Cyphastrea, Pachyseris* are frequent, Faviidae and Fungiidae are also rather frequent. There are 56-70 species of coral that can be met at the reef flat. The width of the zone is 120-300 m. In an ordinary low tide, only a few protrusions reef flat were drained. The community of polyspecific coral settlements may be formed in the area of the inner reef flat. Here funnel-form and lamellar colonies of *A. cytherea, M. aequituberculata* are more frequent, forming up to 50% of substrate coverage. Accumulated coral fragments are inhabited by the following infauna rather thickly: crabs, gastropods, bivalve mollusks, polychetes, and brittle stars - 110-160 spec/m^2 in total with the biomass of 190-200 g/m^2. The biggest population is observed among gastropods *Turbo bruneus* - 30 spec/m^2, and the biggest biomass - among bivalve mollusks *Barbatia bicolorata* - 56 g/m^2. On the reefs at the sandy coasts in such settlements branchy *A. nobilis* and funnel-form *M. danae* usually prevail, with the chalky algae *Amphiroa fragilissima* taking free space among them. Infauna in coral fragments is poor. There are less than 100 species of invertebrates per 1m^2 with the biomass of about 20 g, with the prevalence of crabs *Erotosquilla sp* - 34 spec/m^2 and biomass 10.8 g/m^2. On the reefs with corallogenic substrate, in the prominent outer part of reef flat, there are either monospecific settlements represented by various species of staghorn corals (*A. cytherea, A. nobilis, M. aequituberculata, M. danae* and others), or various combinations of either two of the listed species providing up to 80-100% substrate surface coverage.

Reef-front is a narrow strip of an old reef cemented by chalky algae with numerous caverns. The degree of coverage with live corals is less than 20%. Here there are isolated colonies of massive, massive branchy and crust-incrusting forms of *Psammocora, Pocillopora, Stylophora, Acropora, Favia, Favites, Goniastrea, Porites* being developed. There are thick beds of algae in the niches and caverns, including *Halimeda opuntia* and numerous sea urchins.

The community on the reef slope, as well as on all reefs of Vietnam, differs in the highest degree of substrate coverage with corals (80-100%) and the richest taxonomic diversity of the whole macrobenthos (60-70% of the species composition was studied). It formed on the corallogenic substrate, at the depths of 3 to 4 m, and takes reef parts 40-60 m long, at the distance of 80-200m from the coastal line. No apparent domination of one or two species is observed at all reefs. Peculiarity of reef slope community expressed by considerable prevalence of 3-4 macrobenthos species both by settlement density and size, and by the ability to form zones of monospecific settlements.

In the upper part of the reef slope bioherms (hilly polyspecific settlements) thickly covered with Scleractinian, Alcyonarian and crusted chalky algae are very frequent. Here scleractinian *Pocillopora, Acropora, Montipora, Porites, Goniastrea, Diploastrea, Platygyra, Favia, Favites, Fungia, Herpolitha, Pachyseris, Lobophyllia, Hydnophora* and hydroid *Millepora* are the most frequent (90-115 species in total). *Halimeda, Dictyota, Pocockiella,* and *Halophyla* are the macrophytes represented most numerously on the sandy gaps in channels. The width of the zone reaches 50-70 m, depth drop is 3 to 8 m. The lower part of the slope is more flat. It is a gently sloping part of the bottom coated with the fragments of dead corals (mostly cylindrical) and organogenic dendrite with little impurities of medium-grained sand. Large bioherms have been thickly coated with scleractinian, alcyonarian and single gorgonian. There are large (2-4min the size across) separate massive and massive branchy colonies of *Porites, Diploastrea, Hydrophora, Acropora, Goniastrea, Leptoria,* and *Lobophyllia* between bioherms. Algae were most numerously represented by crusted *Coralinacea, Dictyota* and *Halinicda.* Spots of compact settlements of thin-branchy scleractinian *S. hystrix* met frequently. Both among corals and inside their colonies there are numerous mollusks, among which the most numerous are *A. ventricosa* with the average density of 1.5-2 spec/m^2 and biomass up to 80 g/m^2, and *B. bicolorata* - 2 spec/m^2 and biomass 10.4 g/m^2. Oysters *Lopha cristagalli* are very frequent. Colonies *P. lobata* can be frequently inhabited by large polychaete *Spirobranchus giganteus* (84 species at the biomass of 1127 g for a colony of 15x31x45 cm). On the toe of the reef slope there are mainly lamellar and incrusting forms of scleractinian *Pachyseris, Montipora, Merulina, Echinopora,* sometimes with formation of the zone of *Pachyseris rugosa,* providing 20-38% of the total substrate coverage with corals. Species diversity index in the reef slope zone was the highest and made 9.7.

On the soft soils of pre-reef platform communities of *Malleus malleus+Juncella fragilis* may be formed in front of the reefs, at the rocky coasts on the slimy sands with numerous coral fragments at the depth of 18-21 m. Its basis made up by bivalves *M. malleus* and gorgonian *J. fragilis,* which exceed other macrobenthos species by biomass in dozens of times, though they do not have very high density. Density of *M. malleus* at the different reefs varies from 4,5 to 12,5 spec/m^2 with the *biomass* from 1715 g/m^2 to 2017 g/m^2, and density of *J. fragilis* - from 0,9 to 8,7 spec/m^2 with the biomass from 2 to 313.3 g/m^2. Various polychetes and decapods are a permanent component of the community making up to one third of the total community population.

Quantity of scleractinian here did not exceed 20-25 species, or 10-123% of their composition at the studied reef. Species diversity index made 7.5.

Among the main background species, brittle stars *Amphicmlus laevis* and *Ophiactis savignyi* can always be met in this community. In spite of high population of polychetes (over 100 spec/m^2), their share in the total biomass is almost insignificant and is 0.03-0.05%. There are single bushes of sea grass *H. ovalis* in the sandy gaps. The general picture of community formation is very similar, but their qualitative and quantitative composition varies greatly. The number of mass species varies from 19 to 106. The degree of similarity of different communities keeps within 24.8-37.5, and is slightly less in the communities of soft soil -11.2-24.6. Complexes of scleractinian species have some differences in the degree of similarity, with the difference of extreme values 34.5-41.3. For taxocenoses of mollusks, the degree of similarity is not lower than 32.0 in 50% of communities. Macrophytes with the highest variety in the degree of their species similarity (from 7.1 to 30.7) reduce the degree of similarity of communities. It should be noted that similarity of different neighboring communities at one reef is higher than that of similar communities at different reefs of Vietnam. A high level of similarity of corals is marked between different reefs of South Vietnam and between species compositions Scleractinia of its different regions (Figure 3.7.7).

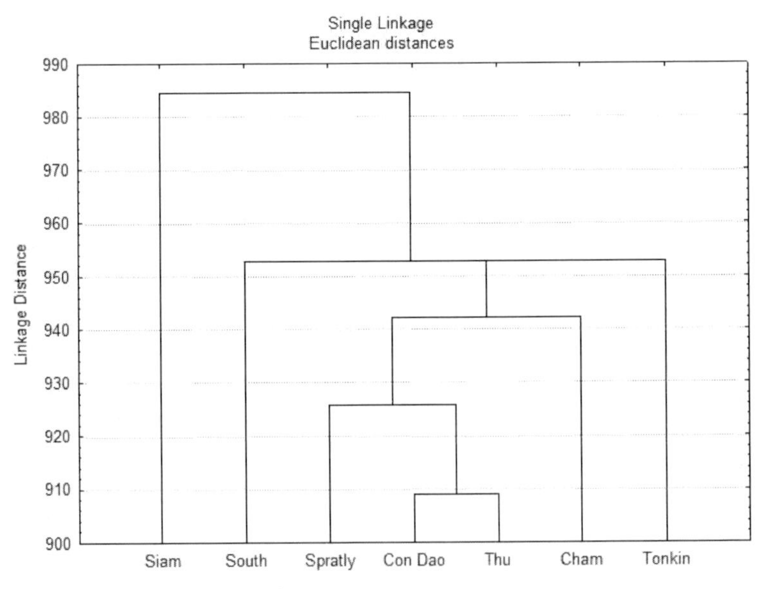

Figure 3.7.7. Dendrogram of similarity of species composition of coral in different regions of Vietnam.

In general, there are 190 to 261 species of scleractinian in the different regions on Vietnamese reefs, which may be compared to the composition of these corals found on the reefs in the open part of the South China Sea. Coral communities at the islands of Con Dao, Thu, Ca Thuik and Lodd are in rather good condition. Both on the structural and structureless reefs, coral communities are characterized by high species diversity.

This region inhabited by over 200 species of scleractinian spread into groups of 90-115 species by separate zones of the reef. Reefs are characterized by high degree of substrate coverage with live corals, and high species diversity of Acroporidae, which proves the optimal conditions for their development and growth. As the structure of reefs has been thoroughly studied before, and has not undergone considerable changes according to our observations, special attention has been paid to taxonomic surveys during the last years. It allowed detecting six species of scleractinian *(Acropora abrolhosensis, A. insignis, A. parilis, Stylophora subseriata, Merulina scabricula, Pachyseris gemmae)*, which were not observed at the reefs of Vietnam before.

It has been noticed before that the species composition of coral fauna, its richness and high diversity, as well as degree of similarity with the coral fauna of the south-western Indo-Pacific allow relating it to the Indonesian-Philippine center of origin of Indo-Pacific tropical corals [32, 38]. Species composition and structure of coral communities of the Vietnamese coral reefs in the open part of the South China Sea do not contradict, but instead make this affirmation even more reliable.

4.2.4. The Gulf of Thailand

Bordering on the Indochina Peninsula in the west and with the Malacca Peninsula in the East, the Gulf of Thailand is the largest gulf of the South China Sea. As distinct from the rest of the Vietnam shelf, which is formed on the submerged periphery of a lithospheric plate, the Gulf of Thailand was formed in the place of a vast depression spreading over 400 km northward and filled with Quaternary deposits over 2500 m thick. The central depression, 60-80 m thick, and the major part of the bottom consist of loose fine-grained deposits underlain with fluvial valleys and smaller alluvial forms of bed facies.

At the time of the winter monsoon, the gulf is entered by one of the branches of the southwestern current, bearing water from the East China and Philippine seas. Uda and Nakao [59] reported that the branch reaches the central part of the gulf, while Sudara [54] states that it declines westward, reaches the Malacca coast, and runs along it finally flowing into the Karimata strait. In this period, circular cyclonic currents and upwelling may arise in the bay, registered via observations of plankton distribution. At the time of the summer monsoon, the waters of the Java Sea moving through the Karimata Strait form the currents of the South China Sea. One of the branches of these currents enters the Gulf of Thailand to give rise to cyclonic and anticyclone circular currents. These run at a speed of some 0.3 knots and involve all the water of the gulf. The southwestern monsoon forces the currents clockwise, while the northwestern, counterclockwise.

The tidal cycle in the gulf is comprised of irregular daily and daily tides with average amplitude of 2.7 m. On the eastern coast of the gulf, regular daily tides are observed, with the amplitude increasing from 1.5 to 3.5 m towards the innermost part of the gulf, where irregular daily tides with amplitude of about 4 m are observed.

In the open part of the South China Sea, the salinity is 32% or above. In various parts of the Gulf of Thailand, it varies between 30.06 and 31.26%. In the innermost part of the gulf, it may drop down to 28% in the rainy seasons [13]. The temperature of the superficial waters ranges from 24 to 30°C being maximum in May-August and minimum in November-February [44]. The water in the gulf is not highly transparent, and the visibility does not exceed 5 m. By their optical properties, the gulf waters vary between the first and fifth coastal water types. In the innermost part of the gulf and in the An Thoi Archipelago, the transparency is minimum corresponding to the fifth coastal water type. In the mouth of the gulf and off the southern coast of Tho Chu Island, the water is most transparent corresponding to the first coastal water type [57].

The reefs of the Gulf of Thailand, situated mostly in its eastern part, develop mainly around archipelagoes and single islands. The islands are relatively high mountainous plateaus with steep marine-cut denudation rock fall slopes. The relief of most islands features steep slopes, well pronounced within the coastal zone both on land and underwater. These geomorphological peculiarities determine the formation of the ruined-rocky submarine relief. The underwater slopes are comprised of boulder-block placers replaced in greater depths with stony and gravel alluvial deposits, which, in turn, transform with depth to sandy and coral deposits with a high content of organogenic detritus [25].

In the 1950s, a number of expeditions were performed along the Vietnam coast by K. Dawydoff [10], a famous Russian researcher of Indochina. In those expeditions, particular attention was drawn to studying reef-building corals including ones off the Namsu and An Thoi archipelagoes, near Cambodian Kong Island and the Ream Promontory, and in the open part of the Gulf of Thailand beginning from the Ream Promontory and ending at the latitude of Tho Chu. Dawydoff composed the most complete list of scleractinian, mainly including ones of South Vietnam. The list comprised 230 species grouped into 51 genera. Some genera contained from 5 to 12 synonymous species names. However, the majority of the species were correctly identified and could be successfully compared to the corals of the other most completely studied regions of the Pacific.

In the 1980s and 1990s, intense studies of the Gulf of Thailand were initiated. High-level joint expeditions were performed focusing on both the reef ecosystem as a whole and the composition and structure of its communities. Using SCUBA diving techniques allowed a considerable increase in the research efficiency and quality. I mean the joint Soviet-Vietnamese expeditions in 1984, 1986, and 1987; the Japan-Thailand expedition in 1984; and the Vietnam expedition performed in cooperation with the WWF in 1994. The results obtained in the course of these expeditions have been presented in a number of reports and publications [22, 25, 49].

In the late 1980s, joint Soviet-Vietnamese expeditions were performed with the purpose of complex study of the coral communities of the Gulf of Thailand. The benthic communities of the Tho Chu Islands and An Thoi archipelago were studied. It is noteworthy that not only the composition and distribution of reef-building corals but also some other characteristics of the coral communities were studied. Thus, new data were obtained on the ecology and distribution of some mass invertebrate species and *Millepora* hydroids, the light effect on the distribution of scleractinian was examined, and other characteristics of the coral reefs and of their inhabitants were studied. In eutrophicated turbid waters of the gulf, corresponding, according to the present classification [19] to the first to fifth coastal water types, a light-dependent distribution of 64 reef-building scleractinian species was revealed. It was also shown that, in the gulf, the vertical distribution of the reef-building scleractinian is confined to a depth of 18-20 m. Another distinctive feature of the Gulf of Thailand coral reefs is the low abundance of scleractinian in the littoral zone. Despite the "depth limitations" of coral reefs and optical heterogeneity of the water, some laws governing the vertical distribution of hermatypic corals and related to the light action were revealed.

The distribution of scleractinian in the gulf, confined to a depth of 18-20 m, was shown to be determined by 8-2% of the incident PAR. This value is close to the light amount limiting the distribution of most zooxanthelae containing corals. The diversity of coral species inhabiting uniformly illuminated bottom areas increases with depth and with an illumination decrease to 3-5% of the incident PAR and decreases again only at the limiting depths. Forty percent of scleractinian were found to inhabit uniformly illuminated bottom areas within the whole vertical distribution range, which is close to the estimates obtained for other Pacific and Atlantic reefs. Eighty-two percent of corals lived at limiting depths at 8-2% of the incident PAR [57]. However, only 8% of the studied species, found to live at less than 3% of the incident PAR, can be considered typically sciophilous. These include *Cycloseris cyclolites, C. patellioformis, Scolymia vitiensis, Leptastrea bewickensis,* and *Trachyphyllia geoffroyi.* The above data suggest the wide adaptive capacity of scleractinian to light intensity and qualitative characteristics.

The composition and distribution of reef-building scleractinian and accompanying mass macrophytes and benthic invertebrates (400-500 species) were studied in the reefs of Tho Chu Island and the An Thoi archipelago. Their zonal distribution was analyzed, and the settlement density and biomass were estimated. Corresponding reef zones were found to be inhabited by the same dominating coral species or groups of them characterized by a certain composition of the macrobenthos. This allowed one to distinguish between several specific communities and reveal their structure, composition, and some features of their development.

The diversity of geomorphological conditions off the An Thoi Archipelago, the continuous inflow of waters from the open part of the South China Sea, stable monsoons, and high levels of water eutrophication result in a rather wide variety of corals (over 170 species) in this region. It is twice as great as the number of coral species in the innermost part of the Gulf of Thailand and is comparable to those off other islands of the gulf, Namsu and Tho Chu. However, it is smaller than the number of scleractinian species off Con Dao and Thu islands, situated in the open part of the gulf. At the same time, compared to the reefs of the open part of the South China Sea, the reefs off the An Thoi Archipelago exhibit a smaller number of acroporids corals, which are replaced by poritids species. The number of *Acropora* and *Montipora* species of Con Dao and Thu islands and the fringing reefs of South Vietnam is 1.5 times as great as that off the An Thoi Archipelago. The reduction of coral species diversity through direct extinction of certain species and a change in the ordinary structure of a reef community, caused by stress reactions of various coral species, are attributed by many researchers to water silting and eutrophication [40, 58].

Under stressful conditions of silted shallows, *Porites* species obviously benefit from their ability to produce a firm bacterial mucous envelope when exposed to contaminated water; to reproduce themselves 1-2 months earlier than the other corals under high eutrophication conditions; to repeatedly produce larvae during an extended reproduction period; and to adapt efficiently to drying, overheat, and desalination. Massive *Porites* corals forming continuous colonies in turbid eutrophicated or silted waters were observed to prevail in many reefs of the Indo-Pacific and Atlantic. The predominance of massive *Porites* corals in almost all reefs of the Gulf of Thailand is caused by their ability not only to survive conditions stressful to many corals but also to maintain a higher organic matter production rate and greater degree of projective coverage of the substrate and to achieve wider species diversity compared to other scleractinian [22, 49, 58].

Figure 3.7.8. A settlement of massive Porites on reefs in the Gulf of Thailand.

The Gulf of Thailand is a shallow gulf with solid rocky-blocky ground formed of thin deposits within a narrow near-shore zone. These peculiarities present some difficulties for further increase of the reef area. Massive *Porites*, constituting one of the major biogenic matter producer groups (100%

occurrence at 40-100% projective coverage of substrate) not only form the reef framework but also contribute much to the enlargement of the reef. Water tides, as well as numerous inhabitants of *Porites* colonies (polychetes, mollusks, crustaceans); cause the erosion of *Porites* colonies. This results in passive colonization of new areas with living fragments of eroded *Porites* colonies [17]. The remnants of perished colonies form a new, now, organogenic substrate, on which *Porites* and other corals and various phyto- and zoobenthos forms settle. Thus, intense production and erosion of the biogenic substrate are a prerequisite for reef growth and formation of fully developed reef-builder communities.

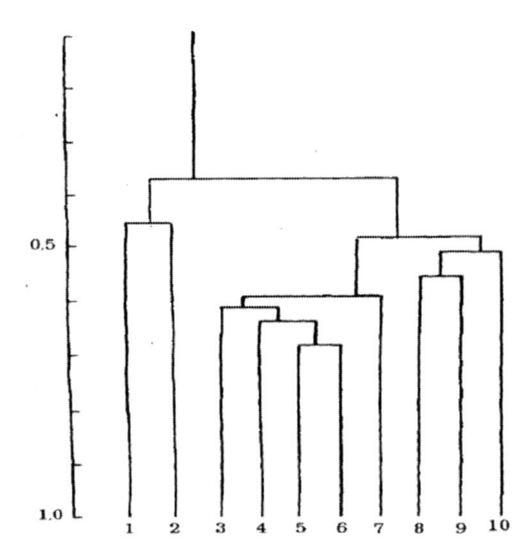

Figure 3.7.9. Similarity dendrogram of the species composition of scleractinian from the Gulf of Thailand and other regions of Vietnam. 1-Bai Tu Long Archipelago (Gulf of Tonkin), 2-Cham Island, 4-Ly Son Island, 3-Hanh Hoa Province, Islands; 5- Thu and Ca Thuik, 6-Con Dao, 7- Tho Chu, 8-Namsu and Khang Khao, 9 -An Thoi Archipelago, 10-South Vietnam. The ordinate, correlation coefficient.

A distinctive feature of the Gulf of Thailand reefs is a weakly pronounced structural and morphological zonality and a small thickness of carbonate deposits of the reef origin, which are common for structural reefs of the Indo-Pacific. In the gulf, the deposits form a thin crust covering boulder-block and rocky substrates, which hardly change the underwater geomorphological relief (see Figure 3.6.2). In spite of this, the composition and structure of the underwater communities exhibit a distinct vertical bionomic zonality, meeting the zonal distribution of environmental factors (substrates, wave regimes, and sedimentation and illumination rates). A similar zonality pattern is observed in both structural and unstructured reefs of the Pacific and Atlantic.

The reefs of the Gulf of Thailand are somewhat similar to encrusting reefs of highly eutrophicated water shallows in that they do not display typical reef zonality and thick deposits of the reef origin. At the same time, most of their features resemble those of "normal" structural reefs of the South China Sea, characterized by a distinct lagoon, reef flat, and other typical zones. The Gulf of Thailand reefs display a high diversity of coral species and pronounced bionomic zonality. The latter is indicative of the presence of two reef zones, the inner autotrophic and the outer heterotrophic, which is typical of classic reefs with distinct morphological zones.

The coral fauna of the Gulf of Thailand is highly similar in species composition to that of the other regions of Vietnam (Figure 3.7.9) and forms a single complex of species of the equatorial reef zone of the Indo-Pacific.

4.3. The Stability and Changes of Coral Communities to Space and Time

Coral reefs of Vietnam are an integral part of Indo-West-Pacific tropic center of origin and diversity of corals [12, 23, 60]. They represent a very productive natural ecosystem with a variety of ecological niches and multiple inhabitants. At the same time, coral reefs serve as a basis for many branches of economy (fishery, construction works, SCUBA diving, etc.). They form a natural barrier against the destructive energy of waves, protecting the coastline from erosion. Coral reef is actually a living barrier, which permanently replaces its destroyed basis due to its growth. Besides, coral reefs through their erosion serve as a main source of sand, replenishing critically eroded coasts. It is especially important for many areas of Vietnamese coastlines with sandy coasts. In 2003 -2005, comparative observations made several reefs of near the city and port of Nha Trang, which we first investigated in 1981. Appreciable changes due to anthropogenic impact have occurred on the reefs that are the nearest to the city. There was a reduction in substrate cover by reef-building corals, a substitution of dominant scleractinian species, and a decrease in the numbers and diversity of common species of corallobionts. The index of species diversity for scleractinian also decreased. The seaweeds *Chnoospora* and *Halimeda* spread into all zones of the reefs. Changes in coral communities on more distant and protected reefs were not so marked. In February 2010, an expedition of the Institute of Marine Biology on the R/V "Akademik Oparin" revisited and surveyed a number of coral reefs of Vietnam, whose species composition and structure had been first described a quarter of century ago. No substantial changes were found in the geomorphological structure of reefs of Tho Chu islands in the Gulf of Siam and Con Dao and Thu islands situated outside the gulf. Marked changes were found in the abundance of some species of associated macrobenthos. The status of coral reefs was described, and the reasons for the observed changes are analyzed (climate, anthropogenic influence, etc. [36]).

Both published and unpublished results of the investigations of Vietnamese reef building corals and reefs performed in the last decades of the twentieth century and first decades twenty-first were analyzed. The state of the art in the study of reef-building scleractinian corals and reefs were presented. The scleractinian fauna of Vietnam showed to be rich in species diversity (350 species of 80 genera) comparable to that of the tropical coral fauna of the Indonesian–Malacca fertile center, from which Indo-Pacific reef-building corals originated. The whole Vietnam coast from the Gulf of Tonkin to the Gulf of Siam is a biogeographically single whole and is a part of the Indo-Polynesian Province of the Indo-Pacific Area.

4.3.1. Description of Coral Reefs

Examined reefs differ by geomorphology, degree of wave effect and connection with open sea, in accordance with which they classified as reefs of open capes and islands, reefs of closed bays, and reefs of channels (Figure 3.8.1).

Near the open coasts of Den, Lon and Mun Islands and near the south-western end of Gom peninsula, under the conditions of active hydrodynamics, coral colonies of comparatively small length (up to 25-100 m off the coast), confined to steep and vertical rocky and stony walls, usually occur. The reefs are characterized by high species diversity of scleractinian (not less than 200 species), and the same high degree of substratum covering (60-100%), by the presence of a big number of young corals settled comparatively not long ago (Figure 3.8.2).

Among the community species, which can be met on the most reefs of Indopacific area, are the most common: *Pocillopora verrucosa, Acropora cytherea, A. florida, A. millepora, A. gemmifera, Montipora hispida,*

M. vietnamensis, Porites lobata, P. cylindrica, Favia maritima, Favites flexuosa, Platygyra daedalea, Leptoria phrygia, Diploastrea heliopora, Goniastrea pectinata, Hydrophora microconos, Lobophylla hemprichii, Symphyllia recta, S. radians, Galaxea fascicularis, Fungia fungites, Sandalolitha robusta, Podabacia crustacea, Merulina ampliata, Pectinia paeonia, Turbinaria peltata and many others. Macroalgae mostly *Padina auastralis* and *Chnoospora implexa*, occurs as shoots mainly in littoral.

From 1982 to 2005, these reefs did not change very much in their species composition and community structure. Diversity of *Acropora*, totaling not less than 30-35 species, remained high as before. The bulk of species diversity, as on the most Indo-Pacific reefs [29, 61], formed by scleractinian from five families:

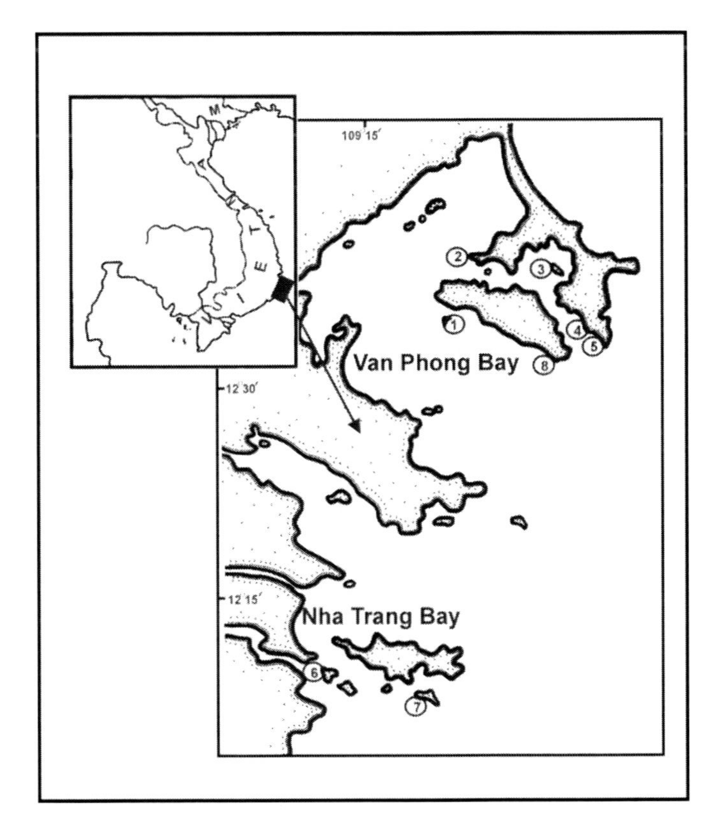

Figure 3.8.1. Location of Van Phong and Nha Trang Bays. Numeric mark number of the transects: 1 –Den Island, 2 – cape Co - co, 3 – Ong Island, 4 – southwest part of peninsula Gom, 5 – cape Thi, 6 – Mju Island, 7 – Mun Island, 8 – southwest part Lom Island.

Acroporidae, Faviidae, Fungiidae, Poritidae and Dendrophylliidae, making more than 60% of their total number. Representatives of five genera – *Acropora* (15-20 species), *Montipora* (10-15), *Porites* (11-13), *Favia* (7-10) and *Fungia* (7-10) – are the most diverse and numerous in coral communities of these reefs. Changes of composition and abundance of macro fauna, associated with corals (sea urchins, holothurians, mollusks), not observed.

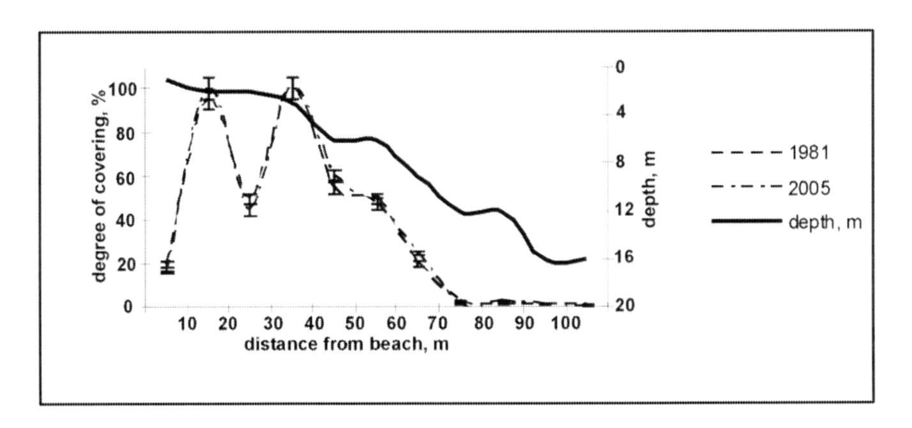

Figure 3.8.2. Variations of a substratum covering of corals on Den Island reef.

Reefs near Ong Island and Co-Co peninsula have a pattern of reefs from closed bays, protected from roughness. A considerable length of lagoon and reef flat zones (sometimes up to 300 m) has lower species diversity (140-160 species), and, overall, a lower degree of substratum covering by corals (not more than 40-60%) is typical for them. As before, dominating of massive *Porites* colonies, formation of microatolls by them, clearly expressed shallow water zone of macrophytes with prevalence of *Sargassum polycystum, P. australis* and *Ch. implexa*, replaced with depth by more abundant *Halimeda opuntia* and *Halophilla ovalis*, are observed here. The most typical are *Porites lobata, P. australiensis, P. rus, P. cylindrica, Favia maritima, Platygyra daedalea, Leptoria phrygia, Acropora millepora, A. monticulosa, A. latistella, Astreopora myriophthalma, Montipora vietnamensis, M. tuberculosa, Fungia fungites, F. paumotensis, F. seychellensis, Podabacea crustacea, Pectinia paeonia* and others.

For the recent twenty years beginning from 1982, the area of substratum covering by corals was substantially reduced on these reefs (down to 10-30%). Small massive coral colonies (5-10 cm), mainly from Faviidae and Poritidae families, began to prevail noticeably in the community structure of the reef slope (not more than 3-5 colonies per sq. m). Silting of substratum, which can be clearly visualized, increased due to the closeness of a big number of mariculture installations. In addition to that, in all areas of intensive mariculture development, there is active aggression of predatory gastropods *Drupella rugosa*, which have density from 8-20 individuals per colony of 10x13x18 cm size to 3000 individuals in some aggregations of branched *Acropora* colonies, against various scleractinian species of *Acropora, Porites* and *Montipora* genera.

A reef near Mju Island is attributed to the channel between the densely populated islands. It is situated in the immediate vicinity to Nha Trang city and port, which is surrounded from every quarter with mariculture farms and tourist complexes. Here the changes in reef community composition are especially obvious due to heavy silting of substratum, corals and other representatives of macrobenthos in the area of the reef slope. In the surrounding waters, values of sedimentation flow are extremely high: 35.3-48.6 $g \cdot m^2 \cdot day^{-1}$. For the past two decades the degree of substratum covering by corals reduced (Figure 3.8.3), number and size of colonies of reef-building scleractinian decreased, and abundance of algae *Halimeda opuntia, H. discoidea* and *Ch. implexa* increased. Species diversity of corals, especially of *Acropora*, was reduced.

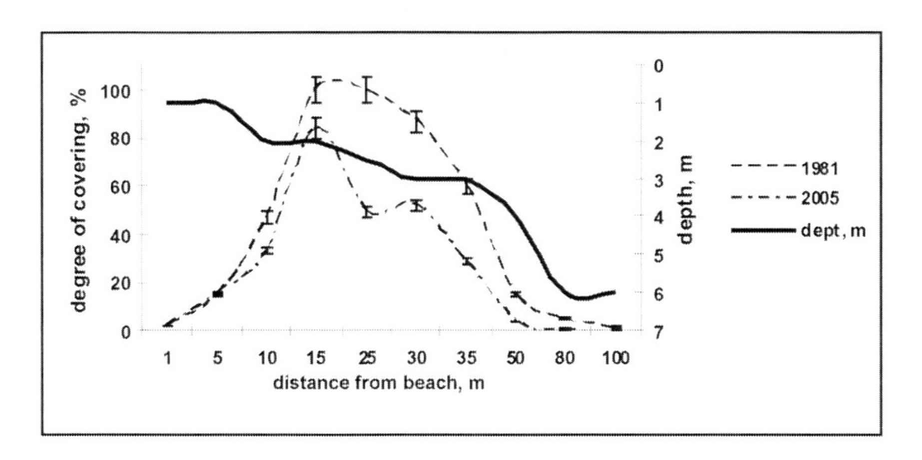

Figure 3.8.3. Variations of a substratum covering of corals on reef Mju Island.

Various species of lamellar and branched *Acropora* and *Montipora*, common here earlier, were considerably replaced by monosettlement of fine-branched *Montipora porites*. Alga *Ch. implexa* settled in all reef zones, occupying actively substratum and space between coral branches, and its covering made 60-75% of substratum area (Figure 3.8.4 a, b). Coral covering of substratum overall rarely exceeds 40-50%. As before, small (2-5 cm) regenerating colonies of scleractinian *Montipora, Porites, Favites,* hydroids *Millepora*, which diversity and abundance, nevertheless dropped 1.5-2 times, can be met on branched debris of dead corals.

Figure 3.8.4. Distribution Chnoospora implexa on Mju Island reef (a), Substratum covering of corals and macrophytes in various zones on Mju Island reef (b). Rectangles correspond to corals, lines – macrophytes a-lagoon, b-reef flat, c-reef slope, d-reef slope base, c-for reef platform.

The changes also affected the macrobenthos accompanying corals. 20-25 years ago sea urchin *Diadema setosum* (not less than 5 individuals per sq. m), holothurians *Holothuria edulis* and *H. atra* (1-2 individuals per sq. m), sea-stars *Linckia laevigata, Culcita novaeguineae, Acantaster planci* (0.1-0.2 individuals per sq. m), mollusks *Atrina vexillum* (up to 0.2 individuals per sq. m), *Tridacna crocea* (0.5 individuals per sq. m), *T. squamosa* (0.1 individuals per sq. m), *Lambis chiragra, L. scorpius, L. lambis, Trochus niloticus, Cypraea tigris, Mauritia arabica* (0.2-0.5 individuals per sq. m) and other invertebrates could be frequently met here. In 2004-2005, only single individuals of *Trochus, Atrina*, sea urchin *Diadema* and holothurians *H. atra* were observed, but the sea star *A. planci* became very common (0.15 individuals per sq. m). Essential changes in species composition and structure of community of scleractinian established only on reef of islands Mju and Mun.

4.3.2. Occurrence Damage of Colonies

From 30 examined scleractinian species (5- *Acropora*, 4 - *Favia,* 4 - *Fungia*, 4– *Porites*, 2 - hydroid Millepora, and to one species another 8 genera) of different growth forms and two hydrocorals 36.7% had signs of decoloration of a part of a colony and/or destruction of its large or small portions. Such portions often looked like scars or cracks, which sometimes were inhabited by drilling mollusks, small macrophytes and Alpheid shrimp (Figure 3.8.4).

On the studies, reefs occurrence frequency of corals with deformation markings of some part of soft tissues or structural anomalies varied very much. It can be clearly seen when comparing deformation frequency of *Porites* colonies. A clear trend can be observed when the reef is located close to an inhabited locality and point to an active area of aquaculture as well as in the case of a large number of damaged coral colonies in the reef community. For example: on the reef near Mju Island, situated near Nha Trang city, a portion of *Porites* colonies with deformation markings is about 63%; near Mun Island, which is considerably distant from the city, it is 21%; near Co-Co Cape (Gom peninsula in Wan Phong Bay), in the area of algae and mollusks cultivation, it is up to 60%, and near the open coast of this peninsula – 37%.

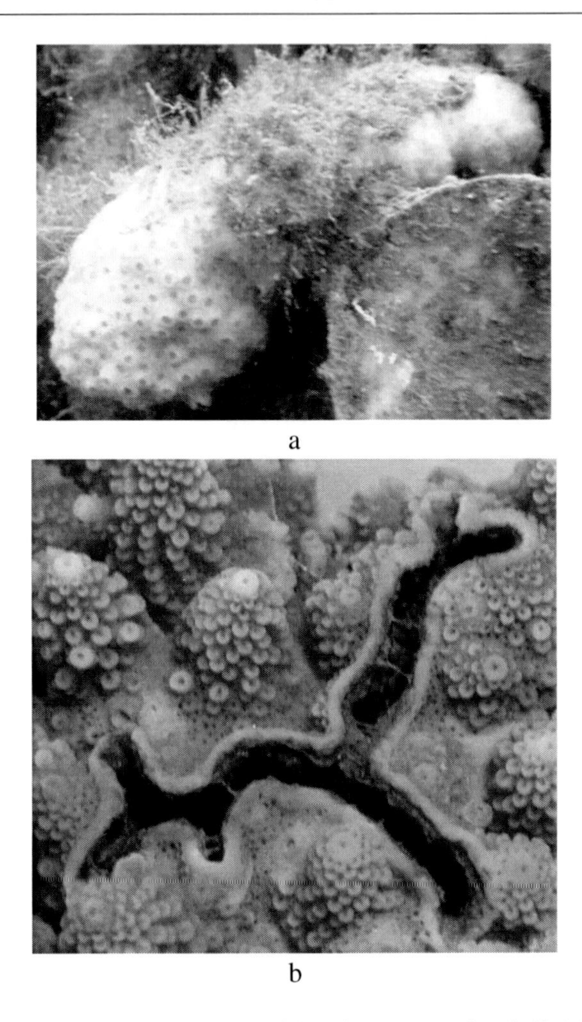

Figure 3.8.4.1. Astreopora ocelata disabled by micro seaweed and silt (a), Acropora gemmifera disabled by shrimp Alpheus deuteropus (b).

4.3.2. The erosive and Mariculture Effect

Erosive consequences of urban development along the coastline, and intensification of mariculture farms in multiple land and island bays greatly increase sedimentation flows and eutrophication of waters of Wan Phong and Nha Trang Bays [2]. Growth of a number of various macroparticles increases water turbidity, reduces photosynthetic resources of reef-building corals and other benthic organisms, and it most probably changes their other physical and biological processes [52]. Connection between sedimentation growth and reduction of species diversity of corals, degree of substratum covering by corals, and low growth rates were shown in many works [3, 58]. Moreover, the increase of the degree of substratum covering by macrophytes was observed under conditions like that [8]. Crude sewage and wastes from mariculture farms usually bring nutrient subsidy and toxic materials in water depth and bottom sediments. The increase in the number of fertilizers and pharmaceuticals near reefs increases the content of chlorofyll. This in turn leads to increased levels of nutrients, and turbidity is the major causes of the significant changes of corals near the coast [14].

Differences found in changes of coral communities on island reefs, outlying inhabited localities and aquaculture areas (Den, Mun), and on reefs exposed to intensive anthropogenic press (Mju, Ong et al.) serve as a striking example of differentiated anthropogenic impact. Species diversity index of scleractinian is more than 2.5 times higher on remote and relatively clean reefs in comparison with that of the reefs exposed to the intensive anthropogenic effect (Figure 3.8.5).

Reefs of islands similar to Mju Island are located in the immediate vicinity to the city and port of Nha Trang. Crowded settlements and tourist complexes are located on its coasts, and multiple mariculture farms operate in its bays. Islands of Mun Island type are farther from the city, and their reefs situated in the reserved protected zone, where there is no population or it does not have numerous representatives of guards and reserve administration. Water clarity near Mju Island and water exchange intensity over coral reefs here is 1.48 times lower than near Mun Island. At the same time, sedimentation flow is 1.3 times greater. Great degree of anthropogenic influence causes growth of eutrophication of waters, surrounding Mju Island, and intensification of substratum silting [31, 43]. Because of these changes, the degree of substratum covering by corals is reduced, and the area of its covering by macrophytes increases. Reduction of diversity of reef-building corals and accompanying mass macrobenthos species takes place. Replacement of *Acropora*, dominated earlier on Mju Island reefs, by fine-branched *Montipora*, having greater total surface area of a colony, can be considered as a possible consequence of high content of dredge deposited in this area (1.3 times higher than in other places).

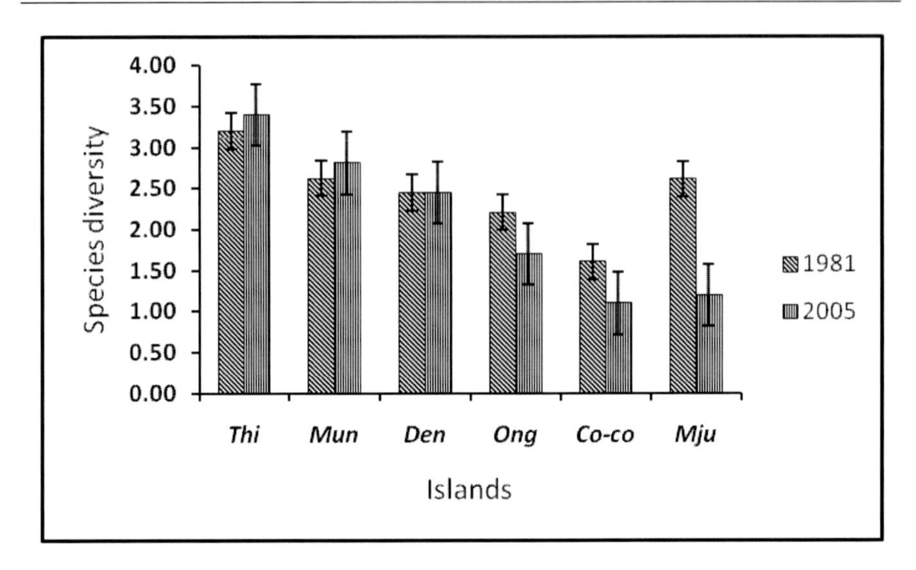

Figure 3.8.5. Variations of a specific diversity on the investigated reefs.

Results of investigation of biological state of separate colonies of different coral species also testify to changes permanently occurring with reef dwellers. Accumulative estimation of representatives of *Porites* genus gives a basis to suggest that various diseases suffered by corals, the number of which steadily increases within recent decades due to discovery of more and more new pathogenic organisms [11, 14], are connected with anthropogenic stress. In the areas of intensive aquaculture in Wan Phong Bay and near Nha Trang city, rather stressed conditions for corals have been formed (high dredge content, low water clarity, intensive sedimentation, multiple cases of mechanical damage of coral colonies, etc.), which facilitate development of various diseases. Such situations are observed on reefs exposed to mass visiting by tourists and in the areas of industrial activity [62].

In many works, which are not necessary to cite here, the level of physical and biological effects disrupting species composition and structure of coral communities have been analyzed. It is universally recognized that the state of coral reefs is noticeably becoming worse on the global level. At present, it is essential to know that we are trying to preserve the diversity of corals on certain reefs, its fish resources, or its ecosystem as a whole. Changes may take place on the level of an individual, population, ecosystem and landscape. Impacts affecting these levels can be short-term and long-term. Short-term impacts can shadow long-term ones. Only long-term monitoring, including single, short-term and long-term natural and anthropogenic impacts, will allow us to estimate stability of coral reef communities and to identify the tendency and reasons for changes. At the same time, it is necessary to strictly observe technologies of marine objects cultivation in the areas of aquaculture to expand the areas and number of preserved and protected zones together with artificial restoration of biological diversity of reef-building scleractinian on reefs. All that will allow us not only to preserve and restore, but also to rationally use the unique ecosystem of Vietnamese reefs.

4.4. CHANGES OF REEF COMMUNITY AFTER TYPHOON

Bays along the coast and islands of Central Vietnam are usually open and sandy, subject to the effect of wind. Their underwater slopes are formed by rubble-lump breakdowns mostly of granite rocks, turning with depth into stony deposits and further into sandy-silty platforms. A big Bung River, bringing thousands of tons of fresh and highly silted water daily, flows into the sea opposite the Ku Lao Cham Islands [56]. This continental outflow can be especially great in the periods of heavy floods and typhoons. Such peculiarity of coastal geomorphology and hydrology of this region has an effect on formation of its sparse reefs, which were thoroughly investigated for the first time by joint Soviet-Vietnamese expeditions in 1984, 1987, and later by the WWF in 1992-1993 [26, 64]. Species composition of corals (more than 200 species), population density of common species of macrobenthos (from 7-10 up to 200 ind/m^2), and degree of substrate cover by corals (75%-100%), formation of continuous settlements *Acropora cytherea* 20%-45% of substrate cover were investigated. It was possible to examine completely satisfactory conditions (optimal temperature and high transparency of the water, the lack of its eutrophication), as well as similarity of these reefs with those of the North and South Vietnam.

In October of 2006, 10 provinces of Central Vietnam went through a powerful Sangshen typhoon. According to data provided by the meteorological center wind, velocity exceeded 133 km/h and sometimes reached 160 km/h. Heavy pouring rains, caused by the natural disaster, resulted in a fast rise of water level in local rivers. In many areas, precipitation levels reached 200 - 230 mm. Erosion of rice fields, coastal and beach areas caused the loss of 80% of mangrove bushes and 95% of coral reefs (based on an unpublished UNU-IAS Seminar Report). As is known, torrents of storm sewage are characterized by great desalination and turbidity due to suspended matter, which is harmful for coral communities and results in their partial or at times complete mortality.

When studying the same reefs at the same locations in 2010, considerable changes in composition and structure of reef communities were found. The community composition changes in this region are of interest for comparative studies with previous data, thereby increasing spatial validity of changes described for reefs of central and southern Vietnam but may be relevant for some areas of the Pacific in which there were similar typhoons.

In 1987, the species diversity on various reefs ranged from 483 to 590 species, among them scleractinian composed 219 species. More than 20 species prevailed: macrophytes *Padina australis, Caulerpa racemosa, Halophyllia ovalis,* alcyonarian *Sinularia dura* and *Sarcophyton trochelioforum*, scleractinian *Acropora cytherea A. robusta, A. humilis, A. digitifera, Montipora aequituberculata, M. hispida, Porites lobata, P. rus, Goniopora stokesi, Pavona decussata, Platygyra daedalia, Diploas-trea heliopora, Galaxea fascicularis* and *Lithophyllum undulatum*, mollusks *Tridacna maxima, Ovula ovum* and *Cyprea arabica*, echinoderms *Diadema setosum, Holothuria atra*, *Linckia laevigata* and *Ophyocoma nigra*. Only one alcyonarian species (*S. dura*) formed mono-specific settlements of colonies on significant areas in various reef zones, and two scleractinian species - *A. cytherea* and *P. lichen* - formed patches of mono-colonies in the reef-flat zone. Formation of several communities could be clearly traced.

The Algal-coral community was formed from the lower level of the intertidal zone to a depth of 2 - 2.5 m. It was characterized by predominant development of algae, subordinate role of scleractinian, and permanent presence of mollusks and echinoderms. Living corals were represented by separate massive and crustacean colonies of *Porites, Goniastrea, Favia, Favites, Galaxea, Hydnophora* and single branched *Pocillopora, Acropora, Seriatopora*. The degree of substrate cover by corals rarely exceeded 10%-15%. Gastropods of the genera *Trochus, Cypraea, Lambis*, echinoderms *Holothuria atra, H. hilla, Stichopus chloronatus* and *Linckia laevigata* permanently occurred among macrophytes and corals.

Reef-flat community. Scleractinian *Acropora cytherea* or *Porites lichen* dominated both by the degree of substrate cover (20%-45%) and by sizes of many-tier colonies reaching 3 m diameter. *Montipora aequituberculata* and *Pachyseris rugosa* were subdominants. Among the other corals, at least 10 species were most frequently encountered. Echinoderms *H. atra, H. edulis* (density up to 4 spec./m^2, biomass 532.2 g/m^2), *Linckia laevigata, C. novaeguineae*, mollusks *Tridacna maxima, Cypraea tigris, C. arabica* were permanent elements of the community. Small patches of the alga *Caulerpa racemosa* were met among coral colonies. Total substrate cover of corals was 60%.

A community of soft corals *Sinularia dura* + *Sarcophyton trochelioforum* was formed at a depth of 4 - 15 m with the width of 20 - 80 m. It was typical for open capes and inlets with rocky and large-lump substrate. Dominating alcyonarian *S. dura* occupied up to 40%-75% of substrate surface. Patches of combined colonies of the corals *A. cytherea* and *M. aequituberculata* with plate trochal colonies up to 3 m diameter were met in shallow parts of the community. Despite domination of soft corals in substrate cover, species diversity was ensured by scleractinian distributed in all part of the community. Acroporidae and branched Poritidae were the most frequent in its shallow part. A variety of species of the 6 genera, as well as *Faviidae* and *Fungiidae*, especially that of the genera *Lithophyllum* and *Cycloseris,* were distributed near the deep part of the reef slope. Echinoderms *Comatula pectinata, L. laevigata, O. nigra, Echinotrix calamaris, E. diadema, D. setosum, H. atra, H. edulis*, and mollusks *O. ovum* and *Lambis lambis* were permanent components of all facies of the soft corals community. *O. nigra* (5 - 7 spec./m^2), *C. pectinata* (10 spec./m^2) and *H. edulis* (4 spec./m^2) were the most frequent.

A community of *Porites australiensis* + *Goniopora somaliensis* with width up to 40 - 60 m was observed from the upper reef slope up to the pre-reef platform. It was typical for open inlets with stony and large-lump organogenic substrate. In the middle of the reef slope a facies of *M. aequituberculata* mono-colony could be formed, which ensured 100% substrate cover together with Poritidae. Massive *Porites* colonies (up to 4.5 m in diameter) with scleractinian *A. cytherea* and *M. Aequituberculata* growing on them were clearly distinguished in the same place. Other corals of the genera *Pachyseris, Echinopora, Montipora, Porites, Micedium, Plesiastrea* and *Turbinaria* occurred in separate colonies or patches of colonies. Faviidae (up to 20 colonies per m^2) with a diameter of 7 - 17 cm were chiefly developed.

Overall, a sufficiently homogenous distribution of corals and accompanying macrobenthos were typical for the reefs of central Vietnam [33]. Zonal distribution of corals could be traced most clearly in open stony reef slopes, at shallow terraces and reef-flats of closed inlets in the presence of organogenic substrate of even small capacity. A similar situation was observed by Dai [7] in Taiwan and by Latypov [24] in North and South Vietnam.

Development of mono-colonies of Acroporidae of the types *A. cytherea* and *M. aequituberculata* in reef-flat zones and on the upper reef slopes, and plate-foveolate colonies of the types *Pachyseris, Mycedium, Echinophyllia* on the lower slope part and on the pre-reef platform, permitted us to integrate the studied reefs with the majority of Indo-Pacific and Atlantic reefs.

Investigation of reefs of Ku Lao Cham Islands (Central Vietnam), conducted by us in the last decade of May of 2010, left a distressing and painful impression. Practically nothing was left from the former richness and beauty of reefs of the reserved islands (Figure 3.9.1).

Figure 3.9.1. Physical destruction and covering Acropora mono-colony with sediments, a- Ku Lao Cham, b – Tcho Chu Islands.

Only 26 scleractinian species were revealed after three days of intensive hydrobiological survey conducted by hydrobiological experts, who spent more than 48 hours underwater and were equipped with modern photo techniques. The species diversity was reduced **tenfold** of mortality of coral reefs along the entire coastline of these islands. There has been a reduction in the coverage of the substrate of Scleractinia and their replacement by Alcyonarian.

A heavy silting of substrate was the first thing noticed when diving. Even vagile animals were covered with a layer of sediments (Figure 3.9.2). Alcyonarian, and only single small scleractinian colonies, mostly covered with silty sediments and various *epibionts,* were typical for the reefs. Representatives of the Acroporidae family, which are mandatory members of all living reefs and usually form the bulk of their species diversity and a high degree of substrate cover by living corals, were completely absent at the reefs. Reductions of species diversity of corals, degrees of substrate cover by corals, and growth delay with the increase of sedimentation were mentioned in many publications [25, 48, 58]. Besides, intensive sedimentation can prevent coral larvae from settling and cause their high mortality after attachment to substrate due to mechanical abrasion. All that leads to fundamental changes in the scleractinian community and to a possible dominance of macrophytes and other coral competitors [50].

Figure 3.9.2. Substrate silting. Motile gastropod Lambis lambis at the left, a dead Faviidae colony at the recto.

Table 1. Cover of hard corals and soft corals before and after the typhoon attack, % covering of substratum

Subject	Algal-coral community	Inner reef-flat	Outer reef-flat	Soft coral community	*Porites* + *Goniopora* community	Reef slope
Macrophytes	10- 40/ 5-7	7-10/ 10	3-5/ 7-10	3-5/ 10-12	3-5/ 10-20	1-5/ 5-7
Scleractinian	10-15/ 0	3-5/ 0.01	45-60/ 0.1	15-20/ 0.01	100/ 0.0	75-100/ 0.01
Alcyonarian	3-7/ 10-18	10/ 10	15/ 45	40-75/ 80-100	0.0/ 100	5-10/ 40 -50

Single scleractinian, found here and there (2 - 4 colonies/m²), are presented mainly by suppressed colonies of massive shape with maximum size of 10 cm in diameter. Massive *Porites* colonies (40%) and representatives of *Favia* and *Goniastrea* (12% each) were the most frequent. Blue corals *Heliopora coerulea* are distributed practically at all transects with a frequency of 7%-10%. In situations like that it is possible to talk about absence of any degree of substrate cover by live corals or about its minimal size, which on the whole, is a small portion of 1%. Colonies of massive Porites up to 1.5 - 2 m in diameter, which can ensure up to 25%-40% degree of substrate cover on some areas of the transect, were found only near Tai Island. Distribution of massive Porites is generally caused by the fact that these corals are able not only to survive in stressful environmental conditions but they reach dominance over the other scleractinian in productivity of organic material, degree of substrate cover and species diversity.

Occurring everywhere, 10 - 15 species of alcyonarian soft corals play the main role in formation of communities on the present reefs of Ku Lao Cham Islands. *Sarcophytum trochelioforum* and *Sinularia dura*, which earlier prevailed among alcyonarian both in occurrence frequency and in the ability to ensure 100% cover of the substrate, dominate. So far inessential but ubiquitous settlement of algae *Caulerpa racemosa, Padina australis, Halophyllia ovalis, Laurencia corymbosa* and *Asparagopsis taxiformis*, which a quarter of a century ago were registered on living reefs only in the coastal zone, where they comprised up to 40%-60% of bottom cover on rocky substrate and dead corals, is now revealed on the reefs. As to macrobenthos, usually accompanying living coral communities and earlier forming mass as assemblages; now single specimens of echinoderms *C. pectinata, L. laevigata, O. nigra, D. setosum, H. atra* and *H. edulis,* mollusks *L. lambis, C. tigris,*

C. arabica and a number of other species occur. It is necessary to mention frequent occurrence of the predatory gastropod *O. ovum* (up to two specimens per one alcyonarian colony), for which soft corals serve as a main foodstuff. Distinct traces of *Ovula* attacks can be clearly seen on many alcyonarian colonies.

Morphological features and vegetative growth of soft corals permit them to endure easier stress situations, including mechanical damage by way of silting. This adaptive strategy of alcyonarian permits them to form mono-dominant communities on many reefs of Central Vietnam, which integrates them, on one hand, with reefs of the Tonkin Gulf and Taiwan with their special hydrodynamic conditions of increased sedimentation, and with reefs of the South Vietnam in the open part of the South China Sea.

This adaptability of alcyonarian can be clearly observed on the studied degraded reefs. Replacement of the former scleractinian-alcyonarian community by a mono-dominant alcyonarian has occurred. Patches of alcyonarian settlements with areas up to several tens of square meters are distributed on all hydrobiological sections. On the southern and Southwestern sides of the Tiem Island, they form compact settlements with 100% substrate cover for hundreds of square meters (Figure 3.9.3). Foraminifers, bryozoans, serpulids and drilling mollusks *Litophaga* spp. gain ground in the forming community together with scleractinian and macrophytes, which designates an ecological type of succession of coelobytic and cryptic organisms [5].

Figure 3.9.3. A mono-colony of the alcyonarian Sarcophytum trocheliophorum.

During the visit to the islands in 2013, it was recorded that the reefs that had been damaged 7 years previously had begun to recover. In the northwestern end of the island, almost intact coral populations were found, which were represented mainly by massive and encrusting–massive colonies of *Porites, Diploastrea, Platygyra, Favia, Goniastrea,* and *Goniopora.*

The *Goniopora* (20% occurrence), *Platygyra* (8.5%), *Pocillopora* (6%), and *Favites* (5.7%) species were widespread in the damaged reef. Moreover, the reproduction of the newly settled scleractinian *Platygyra daedalea* was recorded (Figure 3.9.4). This coral is a hermaphrodite; thus, it does not need a mate for breeding, as the same colony produces both eggs and sperm. It is also important that the larvae settle while undergoing metamorphosis within 3–4 days after fertilization, while in the absence of a substrate their planula retains the ability to settle for 3 months.

Figure 3.9.4. The reproduction of the newly settled scleractinian Platygyra daedalea.

In October 2009, the Vietnam landfall of typhoon "Ketsana," which reached gusts of wind up to 165 km/h corresponded to category 2 on a scale Saffir-Simpson Hurricane. The typhoon was accompanied by heavy rainfall (about 200 mm of rain), and created waves more than two meters tall. After comparable in strength typhoon corals are destroyed at depths greater than 12 m - 60% to 80% - between 12 m and 30 m and 100% - beyond 35 m, whereas earlier living coral coverage ranged from 60 to 75% in these zones. Most of the reefs of the island Tho Chu (Thailand Gulf) were virtually destroyed, when the coastal zone at depth of 5.3 m, 40% of its corals were destroyed; at depth of 5-8 m in the settlement of *Acropora*–100%; at depth of 8-12 m – 60%; 5.3 m, 40% of its corals were destroyed; at depth of 5 - 8 m in the settlement of Acropora–100%; at depth of 8-12 m–60%; at depth of 12 m–10%. In this regard, it is appropriate to go back briefly to the information about the state of these reefs a quarter century ago. The typhoon caused significant damage of the rich scleractinian species, which declined by almost a third (from 275 to 95 species). Substrate coral cover was decreased by 2-3 times as well as an index of species diversity (Figure 3.9.5).

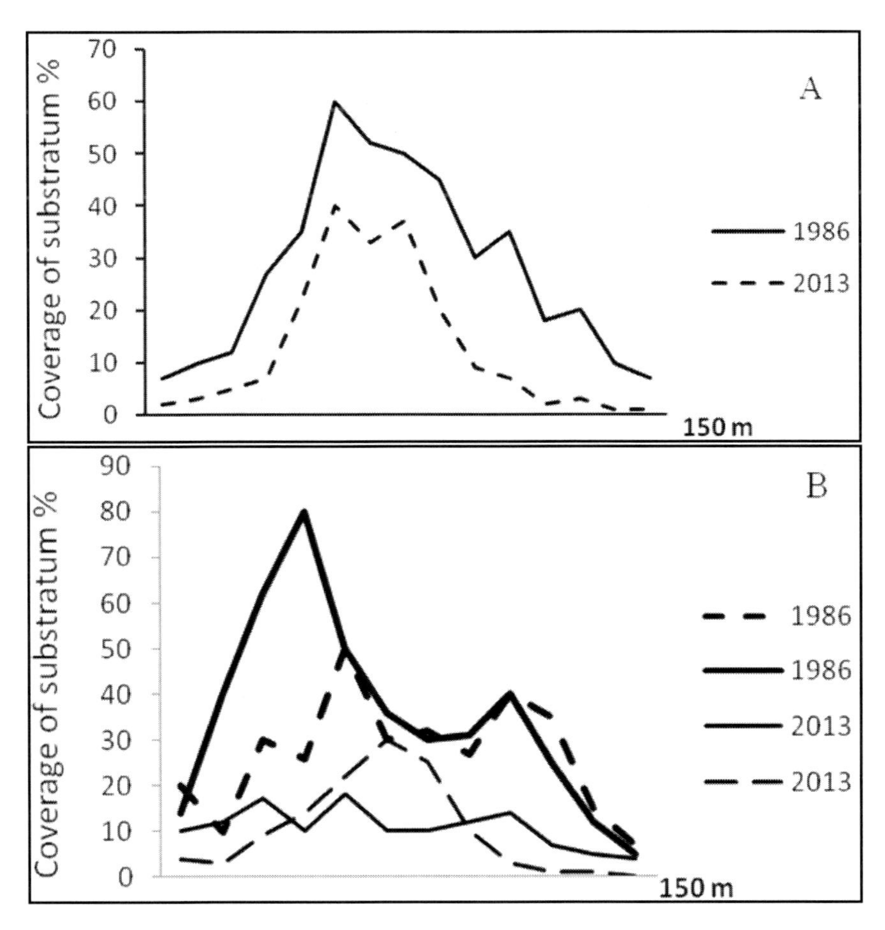

Figure 3.9.5. Coverage of the substrate corals: A - Common, B - Branched (solid line) and massive (dashed line) colonies.

There are opportunities to restore reefs through the replenishment of coral, larvae, and aquatic organisms associated with other reef communities. To the south of the island (in the distance of one mile 9°16'N, 103°21'E), there is a bank with depths ranging from 4 to 13 meters from the optimally developed noticeably damaged by typhoon coral community. There's the usual form for the settlement of various coral reefs more than 100 species, many common, at least 20 species *Acropora*, and related corallobionts typical Vietnamese optimal reef community.

Organisms with calcareous skeletons, especially crustacean calcareous algae, play the main role at the initial succession stages in formation of a stable substrate for settling and growth of hermatypic corals. In 2 - 3 years after calcareous substrate formation, the dominance of reef building coral colonies become appreciable, and only in 4 - 6 years after the beginning of a new reef colonization is the appearance of vast colonies of *Acropora* corals possible [47, 5]. However, there is hope for restoration of the studied reefs, as we managed to find single, very small, young, settled a new colonies of *Pocillopora verrucosa* - one from mass scleractinian species on all Indo-Pacific reefs.

4.5. Rehabilitation of Coral Reef

4.5.1. Concept of Rehabilitation of Corals

According to the Global Coral Reef Monitoring Network (GCRMN) organization, 20% of coral reefs have been destroyed by human activity. The main causes of coral reef devastation are general global problems: climate warming and increasing acidity of oceanic waters through carbon dioxide emissions to the atmosphere, increased eutrophication and sedimentation of coastal waters, reducing fish stocks and the barbaric attitude of the residents of coastal areas to sea inhabitants. A very harmful impact on the coral reefs is caused by collecting corals, shells and other tropical fauna. The world's annual income from coral sales is estimated at 30 million U.S. dollars.

The world scientific community and civil society are concerned with these sad, even threatening circumstances. Special workshops are held; the results of research on the problems of conservation and restoration of coral reefs and the results of experiments on artificial breeding of reef-building corals are regularly published [6, 30, 52]. It is shown that the transplantation of coral fragments is the most successful method of restoring reefs. Under favorable conditions, coral fragments attach to the substrate and, when recovered to the size of a colony reproduce sexually. Earlier, a detailed analysis was performed on the theoretical works and the experimental data on artificial cultivation of corals, and the "framework" method was recognized as most appropriate for Vietnam; it has given good results. It is known that coral fragments are more likely than the larvae to survive in loose (mobile) soils, owing to their larger sizes and elevated localization above the substrate [16, 25].

Reproduction of coral communities through fragmentation of colonies seems to be a reasonable and effective method. But before intervention into the reef ecosystem (in order to assess all the critical situations connected with the use of the method) we shall experimentally determine the procedures that would allow us to obtain the greatest number of survived transplanted fragments at the least damage of live colonies. For that, we must elucidate the following important questions: which coral species give a larger number of viable fragments; from what parts of the donor colony shall we select the fragments for transplantation; what size should the transplanted fragments be; what is the preferred orientation of the coral fragments at transplantation; what should be the depth of planting; and whether the bottom should be cleaned from algal growths.

4.5.2. Materials and Methods

Based on an agreement with the Institute of Technology and Applied Research of the Vietnamese Academy of Science and Technology and "Sanest" Company, in April 2010 we installed three 1 × 2 m plastic frames, elevated 50-70 cm above the bottom, near the protected Hon Nye Island. Coral fragments (115 pieces) of 14 species of the genera *Acropora, Isopora, Pocillopora*, and *Porites* were attached to the frames. The frames were installed at a distance of 70-80 m from the water edge at a depth of 3-4 m in the sand-corallogenic hollows among dense growths of reef-building coral colonies. Transplantation of fragments was performed without removing the corals from water.

Fragments with 2 to 17 branches were taken out of the periphery of the donor colony, placed in individual plastic bags and transported from the nearest assemblages of donor corals to an installation at a distance of 30-50 m, and depth of 2-3 m. The fragment length was measured with a slide gage, and three size groups were identified: 4 -7, 11-12 and 20-21 cm. Coral pieces were attached to the frames with a copper wire in a plastic sheath; their contact with the framework was possibly avoided. The survival and growth rates of coral fragments were investigated in terms of the coral species, the fragment size and its orientation at attachment, as well as of the season of transplantation. The installation with coral fragments was raised above the bottom to prevent the covering with sedimentation and a possible attack of the predatory gastropod *Drupella rugosa*. The state of the facilities and the attached coral fragments was checked in a week. The first results were recorded six months from the start of the experiment.

4.5.3. Results of Experiment

All survived transplanted fragments recovered and formed new colonies with numerous branches. No dependence was detected among the number of survived fragments, their orientation at planting and the season of transplantation. About 60% of the fragments had overgrown the wire connecting them to the frame and within 4-6 months formed basal attachments to horizontal bars of the installation, using them as a substrate. A year after the transplantation, all successfully survived fragments were found attached to the frame of the installations and had the size of a small colony. The survival of fragments was 100-86.2% and depended on the coral species, fragment size and duration of the experiment (Figure 3.10.1).

Figure 3.10.1. Experimental industrial installation for artificial cultivation of corals.

In the first size group, 20-30% of transplanted fragments died within 8-10 months of beginning the transplantation. Small- and medium-sized fragments of *Acropora valida* and *A. microphthalma* showed the lowest survival. Medium- and large-sized fragments of the species *Acropora valida, A. valenciennesi, A. florida, A. gemmifera,* and *Pocillopora verrucosa*, *Porites attenuata* and *P. cylindrica* regenerated most successfully. Fragments of *Isopora palifera* had 100% survival irrespective of orientation at transplantation. Within 2-2.5 months of transplantation, new branches formed on the branch surfaces of all the survived coral species, including the bottom surfaces damaged at separation from the donor colonies. The linear growth increment of coral fragments of various species was 30-160 mm. This parameter depended on the species and size of the fragment. Morphologically different fragments of *Acropora valida, A. valenciennesi, A. formosa, P. attenuata* and *P. verrucosa* were characterized by different growth rates. A high linear growth increment was typical of the most extensively branched fragments of *A. valida* and *P. attenuata* (Table 2, Figure 3.10.2). Large fragments were also characterized by a higher growth rate.

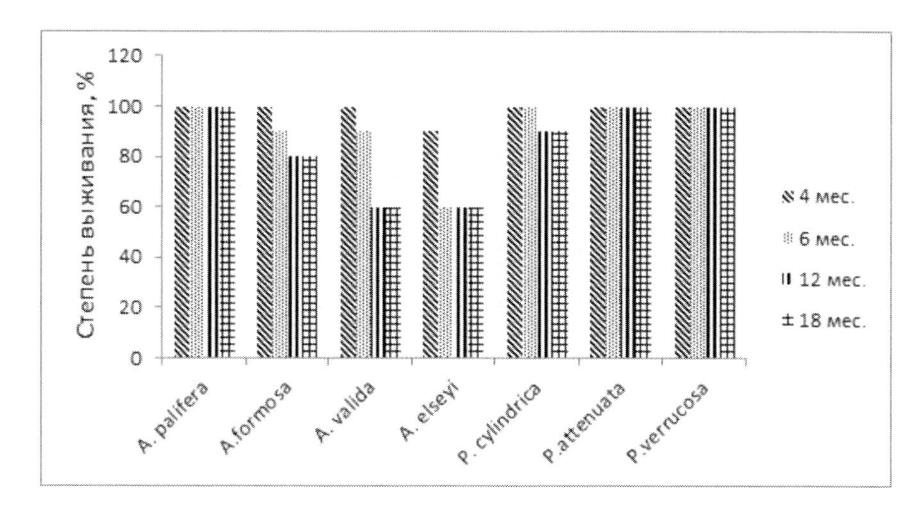

Figure 3.10.2. Survival of transplanted fragments of various coral species.

The more ramified the donor colonies, the more new branches appeared on the transplanted fragments. During the first six months of 2010, the size of fragments and the number of new branches on them generally increased by 150-165%, and in a year by 210-275% (Figure 3.10.3).

Figure 3.10.3. Linear growth increment of coral fragments and an increase in the number of branches in the colony of Porites attenuata. (a) transplanted fragment; (б) upon 6 months; (в) upon 12 months.

Table 2. Number of branching's on the transplanted fragment in terms of species and size of coral fragments in the beginning and the end of the experiment

Species	Period of observations (month)	Number of branches on the fragment	
		Beginning of the experiment	End of the experiment
Acropora valida	12	12	57
A. valenciennesi	12	7	47
A. microphthalma	18	4	37
A. formosa	18	17	42
A. robusta	12	4	16
A. elseyi	12	12	83
A. cerealis	12	12	79
Isopora palifera	18	4	7
Pocillopra verrucosa	12	20	40
P. eydouxi	12	12	26
P. woodjonesi	12	5	15
Porites cylindrica	12	5	65
P. nigrescens	12	7	18
P. attenuata	12	32	75

At extension of the cultivation period from one year to eighteen months, the coral fragments increased 1.2-1.5 times. The successful growth of transplanted fragments and the formation of large colonies contributed to invasion of the damselfish *Dascyllus reticulatus* (Pomacentridae) into the new coral assemblages (Figure 3.10.4).

Figure 3.10.4. Put the fish in an artificial coral Dascyllus reticulatus does settlement.

This coral reef fish demonstrates a pronounced homing behavior in the formed coral reefs similar to that in the natural reef, where adult fish usually live in groups in thickets of branched corals and seldom separate more than a distance of 1 m.

Chapter 5

CONCLUSION

The conducted experiments have shown that cultivation of coral fragments on artificial facilities can be effective in different reef parts. Large fragments of the studied coral species survived better and had higher growth rates. They formed the greatest number of new branches and built large colonies, which is consistent with the results obtained earlier [53, 30, 35]. It is known that large coral colonies grown from larger fragments also have the greatest reproductive success [42].

Growth of corals from colony fragments is an important natural process, at least, in corals with branched colonies. Under natural conditions, the broken fragments of colonies first "anchor" on the bottom and then attach to the substrate through regeneration and growth of the soft tissues and skeleton. The results obtained in our experiments agree well with the data reported by Okubo with co-authors [42], who believe that attachment of fragments to the substrate is a prerequisite for the successful completion of a long process of transplantation. In the experiment of 2010-2011, all the survived fragments became rooted to the frame and attached with their base parts to the horizontal bars of the installation. A direct proof of the formation of the coral settlement is its invasion by reef fish, and an indirect proof is the settlement and growth of sea squirts on this artificial substrate.

Thus, the experiments carried out in 2010-2011 on the coral reefs of Vietnam have confirmed that successful survival of coral fragments under natural conditions of coral reef is possible and depends on the two main factors: the coral species and the size of coral fragment to be planted. Relatively high growth rates of fragments of all coral species can probably be explained by their transplantation into a well-lit environment less populated with other macrobenthic organisms. Installation of the experimental facilities over the bottom keeps them from getting buried with sandy sediments.

The data obtained in our experiments can be used for restoring natural coral settlements or for cultivation of corals for aquariums and oceanariums. A chain of facilities installed on the bottom of sandy areas along or around a reef contributes to increasing the area of the reef within two to four years, which helps protect the shore from wave action.

We damaged the reefs. We must understand that without our help, they will cope. However, we now know that reefs can respond to active human rescue actions positively, so they will hopefully provide benefits and joy for future generations.

REFERENCES

[1] An, N. T.: Biological productivity of Vietnam marine waters. Monography on Vietnam seas. Hanoi: Sci. and Techn. Publ. House, 63-69 (1994).

[2] An, N. T., Son, V. D., Thu, P. M., and Huan, N. H.: Tracing Sediment Transport and Bed Regime in Nha Trang Bay. *Coll. Mar. Res. Works*. 10, 63–69 (2000).

[3] Andres, N. G., Witman, J. D.: Trends in Community Structure on a Jamaican Reef. *Mar. Ecol. Prog. Series*. 118, 305-310. (1985).

[4] Barnes, R., Hughes, R.: An Introduction to Marine Ecology; Third Edition. Malden, MA: *Blackwell Scie. Inc.* 117-141. (1999).

[5] Choi, D. R.: Ecological Succession of Reef Cavity-Dwellers (Coelobites) in Coral Rubble. *Bull. Mar. Scie*. 35(1), 72-79 (1984).

[6] Cox, E. F.: Fragmentation in the Hawaiian Coral Montipora verrucosa. In Proceeding. 7th International Coral Reef Symposium, 513–516 (1992).

[7] Dai, C. F.: Patterns of Coral Distribution and Benthic Space Partitioning on the Fringing Reefs of Southern Taiwan. *Mar. Ecol.* 14(3), 185-204 (1993).

[8] Dai, C. F.: Dynamics of Coral Communities, Biodiversity and the Dynamics of Ecosystems. *DIWPA Ser.* 1, 247–265 (1996).

[9] Davidson, O. G. The enchanted braid: coming to terms with nature on the coral reef. John Wiley & Sons, Inc., New York. (1998).

[10] Dawydoff, C.: Contribution a l'etude des invertebres de la faune marine benthiques de l'Indochine. *Bull. Biol. France Belg. Suppl.* 37, 1-158. (1952).

[11] Denner, E. B. M., Smith, G., Busse, H. J. et al.: *Aurantimonas coralicida* gen. nov., sp. nov., the causative agent of white plague type II on

Caribbean scleractinian corals. *Intern. J. Sys. Evol. Microbiol.* 53, 1115-1122 (2003).

[12] Ekman, S.: *Zoogeography of the Sea*, London: Sidgwik And Jackson, 1953.

[13] Emery, E.O., Niino, I.: Sediments of the Gulf of Thailand and Adjacent Continental Shelf, *Geol. Soc. Amer. Bull.* 74, 541-554 (1963).

[14] Fabricius, K. E.: Effects of irradiance, flow, and colony pigmentation on the temperature microenvironment around corals. Implications for coral bleaching? *Limn. and Oceanog.* 51, 30–37 (2006).

[15] Goreau, T. F., Goreau, N. I., Goreau T. J.: Corals and Coral Reefs. *Scientific American*. (1979).

[16] Hall, V.R.: Interspecific Differences in the Regeneration of Artificial Injuries on Scleractinian Corals. *J. Exp. Mar. Biol. Ecol.* 212, 9–23 (1997).

[17] Highsmith, R.C.: Passive Colonisation and Asexual Colony Multiplication in the Massive Coral *Porite lutea* Milne Edwards and Haime. *J. Exp. Mar. Biol. Ecol.* 47, 55-67 (1980).

[18] Highsmith, R.C.: Reproduction by fragmentation in corals. M*ar. Ecol. Progr. Ser.* 7, 207-226 (1982).

[19] Jerlov, N.G.: Optical oceanography, Elsevier: Amsterdam, 1968. Translated under the title Opticheskaya okeanografiya. Moscow: Mir, (1970).

[20] Jones, O.A., Endean, R.: Biology and geology of coral reefs, Vol. II. Nutrition of corals.. Academic Press, New York. 77-116 (1973).

[21] Ken, L.V.: Stony Corals and the Coral Reefs of Catba Islands, Marine Environment and Resources. Hanoi: Sci. Tech. Publ. House. 144-151 (1991).

[22] Latypov, Yu. Ya.: Coral Communities of the Namsu Islands (Gulf of Siam, South China Sea). Mar. Ecol. Progr. Ser. 29, 261-270 (1986).

[23] Latypov, Yu. Ya.: Community Structure of Scleractinian Reefs in the Baitylong Archipelago (South China Sea). *Asian Mar. Biol.* 12, 27-37 (1995).

[24] Latypov, Yu. Ya.: Benthic Communities of Coral Reefs of Tho Chu Island (Gulf of Siam, South China Sea*). Biol. Morya.* 25(3) 233-241 (1999).

[25] Latypov, Yu. Ya.: Macrobenthos Communities on Reefs of the An Thoi Archipelago of the South China Sea. *Russian J. Mar. Biol.* 26(1), 18-26 (2000).

[26] Latypov, Yu. Ya.: Coral Reef Communities of the Central Vietnam. *Biol. Morya*. 27, 238-241 (2001).
[27] Latypov, Yu. Ya.: Reef-Building Corals and Reefs of Vietnam: 1. The Gulf of Thailand. *Russian J. Mar. Biol.* 29(1), S22-S33 (2003).
[28] Latypov, Yu. Ya.: Reef-Building Corals and Reefs of Vietnam: 2. The Gulf of Tonkin. *Russian J. Mar. Biol.* 29,(1), S34-S45 (2003).
[29] Latypov, Yu. Ya.: Reef-Building Corals of Vietnam as a Part of the Indo-Pacific Reef Ecosystem. *Russian J. Mar. Biol.* 31, S34-S40 (2005).
[30] Latypov, Yu. Ya.: Variation in Composition and Structure of Coral communities of Mju and Mun Islands (Nha Trang Bay, South Chinese Sea). *Biol. Morya*. 32(5), 326–332 (2006).
[31] Latypov, Yu. Ya.: Coral Reefs of Vietnam. *Nauka*, Moscow, (2007).
[32] Latypov, Yu. Ya.: Species Composition and Distribution of Scleractinians on the Reefs of the Seychelles Islands. *Russian J. Mar. Biol.* 35(6), 454–462 (2009).
[33] Latypov, Yu. Ya.: Scleractinian Corals and Reefs of Vietnam as a Part of the Pacific Reef Ecosystem. *Open J. Mar. Scie.* 1, 50-68 (2011).
[34] Latypov, Yu. Ya. Spratly Archipelago as a Potential Reserve Recovery of Biodiversity in Coastal and Island Reefs of Vietnam. *Mar. Sci.* 2(4), 34-38 (2012).
[35] Latypov, Yu. Ya.: Barrier and Platform Reefs of the Vietnamese Coast of the South China Sea. *Intern. J. Mar. Sci.* 3(4), 23-32 (2013).
[36] Latypov, Yu. Ya., Selin, N.I.: Current status of coral reefs of islands in the Gulf of Siam and southern Vietnam. *Russian J. Mar. Biol.* 37, 246-253 (2011).
[37] Latypov, Yu. Ya., Selin, N.I.: Changes of Reef Community near Ku Lao Cham Islands (South China Sea) after Sangshen Typhoon. *Amer. J. Climate Change*. 1, 41-47 (2012).
[38] Latypov, Yu. Ya., Long, P.Q.: The common hard corals of Vietnam. Ministry of Agriculture and Rural Development, Hanoi. (2010).
[39] Loi, T. N.: Peuplements Animaux et Vegetaux du Substrat des Intertidal de la Baie de Nha Trang (Vietnam). *Mem. Inst. Oceanog*. 11, 1-236 (1967).
[40] Loya, Y.: Effects of Water Turbidity and Sedimentation on the Community Structure of Puerto Rican Corals. *Bull. Mar. Sci.* 26(4), 450-466 (1976).
[41] Malyutin, A.N., Latypov, Yu. Ya.: Distribution of Corals and Biogeographic Zonation of the Shelf of Vietnam. *Biol. Morya*. 4, 26-35 (1991).

[42] Okubo, N., Taniguchi, H., Motokawa, T.: Successful Methods for Transplanting Fragments of *Acropora formosa* and *Acropora hyacinthus. Coral Reefs.* 24, 333–342 (2005).

[43] Pastorok, R. A., Bilyard, G. R.: Effects of sewage pollution on coral-reef communities. *Mar. Ecol. Prog. Ser.* 21, 175-189 (1985).

[44] Pham, H.H.: Vegetation of Phu Quoc Island (Thuc Vat Dao Phu Quoc). *Ho Chi Minh Publishing Co.* 128-156. (1985).

[45] Picard, J.: Essai de classement des grands types de peuplements marins benthiques tropicaux, d'apres les observations effectuees dans les parages de Tulear (Sud Quest de Madagascar). *Rec. Trav. Stn. Mar. Endoume, Fasc. Hors Ser.* 3–24 (1967).

[46] Pichon, M.: Dynamic of Benthic Communities in the Coral Reefs of Tule'ar (Madagaskar): Succession and Transformation of the Biotops Trough Reef Tract Evolution. In: Proceeding of 2nd International Coral Reef Symposium, vol. 2, 55-68 (1974).

[47] Preobrazhensky, B.V., Latypov, Yu. Ya.: Regeneration Processes in Coral Reef Ecosystems, Biologiya koralovykh rifov. Morphologiya, systematika, ekologiya (Biology of Coral Reefs. Morphology, Systematics, Ecology), M.: Nauka. 5–12 (1980) in Russian.

[48] Sakai K. and Nishihira M., Immediate Effect of Terrestrial Runoff on a Coral Community near a River Mouth in Okinawa, Galaxea, Vol. 10, 1991, pp. 125-134.

[49] Sakai, K., Yeemin, T., Svidvong, A., Nishihira, M.: Distribution and Community Structure of Hermatypic Corals in the Sichang Islands, Inner Part of the Gulf of Thailand. Galaxea. 5(1), 27-74 (1986).

[50] Salvat, B.: Dredging in Coral Reefs, In: B. Salvat, Ed., Human Impact on Coral Reefs: Facts and Recommendations, Museum National D'histoire Naturelle et École Pratique des Hautes études, Antenne de Tahiti & Centre de l'Environnement, California. 165-184 (1987).

[51] Sèrene, R.: Inventaires des invertebres marine de l'Indochine. Inst. Oceanogr. Indochine. 30, 3-83 (1937).

[52] Soong, K.T., Chen, T.: Coral Transplantation: Regeneration and Growth of Acropora Fragments in a Nursery. Restor. Ecol. 11(1), 1–10 (2003)

[53] Smith, L.D., Hughes, T.P.: An Experimental Assessment of Survival, Re-Attachment and Fecundity of Coral Fragments. *J. Exp. Mar. Biol. Ecol.* 235 147–164 (1999).

[54] Sudara, S.: The Distribution of Planctonic Hyperiids (Crustacea, Amphipoda) in the South China Sea and the relationship to the

Distribution in the Gulf of Thailand. In Proceeding III CSK Symposium, Bangkok, Thailand. Hydrogr. *Dep. Press*. 293-314 (1973).

[55] Sumich, J. L.: An Introduction to the Biology of Marine Life; Sixth Edition. Dubuque, IA: Wm. C. Brown. 255-269 (1996).

[56] Thanh, T. D.: Change in Environment and Ecosystems Relative to the Land-Sea Interaction in the Vietnam Coastal Zone. Reports EALOICZ Workshop, Qingdao. 1-10 (1999).

[57] Titlyanov, E.A., Latypov, Y.Y.: Light-Dependence in Scleractinian Distribution in the Sublittoral Zone of South China Sea Islands. *Coral Reefs*. 10, 133-138 (1991).

[58] Tomascik, T., Sander, F.: Effect of Eutrophication on Reef-Building Corals. 3. Reproduction of the Reef-Building Coral *Porites porites*. *Mar. Biol*. 94(1), 77-94 (1987).

[59] Uda, I., Nakao, T.: Water Masses and Currents in the South China Sea and Their Seasonal Changes. In Proceeding III CSK Symposium. Bangkok, Thailand. *Hydrogr. Dep. Press*. 161-168 (1973).

[60] Veron, J. E. N.: Corals in Space and Time: The *Biogeography* and Evolution of the Scleractinia. Cornell University Press, New York, 1995.

[61] Veron, J.E.N. and Hodgson, G., Annotated Checklist of the Hermatypic Corals of the Philippines, *Pacif. Sci.* 43, 234-287 (1989).

[62] Vo, S.T., De Vantier, L., Long, N.V., et al.: Coral Reefs of the Hon Mun Marine Protected Area, Nha Trang Bay, Vietnam, 2002: Species Composition, Community Structure, Status and Management Recommendation. In Proceeding of Science Conference. Bien Dong-2002, Nhatrang, Vietnam. 650–690 (2002).

[63] Vo, S.T., Yet, N.H., Alino, P.M.: Coral and Coral Reefs in the North of Spratly Archipelago – The Results of RP-VN JOMSRE-SCS 1996. In Proceeding Science Conference. RP-VN JOMSRE-SCS 96, Hanoi, 22–23 April. 87–101 (1997).

[64] WWF.: Vietnam Marine Conservation Southern Survey Team. Survey Report on the Biodiversity Resource Utilization and Conservation Potential of Coto Islands, Quangninh Province, N. Vietnam, Gland, Switzerland, (1994).

[65] Yet, N.H.: Thánm phán loái hô cúng và cáu trúc ran san hô Dâo Thuyên Chài (Quân Dâo Truòng Sa), Tai nguyên và mô i truong biên. Hanoi, 1(IV), 299–313 (1997).

[66] Yet, N.H., Ken, L.V.: Some Data on Species Composition and Distribution of Scleractinian Corals in Ha Long Bay. *J. Biol.* (Hanoi). 18, 7-13 (1996).

GLOSSARY

Abiotic (factor) - describes a physico-chemical factor (in contrast to a biotic or biological factor) of the environment in which an organism lives.

Alcyonaceans - animals belonging to the phylum Cnidaha, with eight tentacles, generally living in colonies, often called soft corals.

Anthropogenic - resulting from human activities.

Aragonite - a natural form of calcium carbonate.

Azooxanthellate corals - corals that do not have zooxanthelae.

Benthic - living on or near the bottom of the ocean. The benthos is the group of organisms living on or near the bottom of the ocean.

Biodiversity - the term has recently acquired many meanings, but can be considered synonymous with 'systematic diversity.'

Biogeography - the study of the geographic distribution of life and the reasons for it. In practice, biogeography is divisible into observations of distributions and explanations of those observations.

Biomass - total mass of living organisms, population explosion of an animal or plant.

Biotope - a geographic area that is under the influence of environmental parameters, the dominant characteristics of which are homogeneous. Biotopes are generally the smallest ecological units that can be delimited by convenient boundaries and which are characterized by their biota.

Branching colonies - any growth-form where branches are formed.

Caespitose - a descriptive term for branches, which interlock similarly in three dimensions. Applied only to the genus *Acropora.*

Calcite - a natural form of calcium carbonate.

Calcitic skeletons - skeletons primarily composed of the calcite form of calcium carbonate. All molluscs have calcite skeletons.

Calice - the upper surface of a corallite to which the soft parts of an individual polyp are attached.

Cerioid corals - massive corals that have corallites sharing common walls.

Climax - ideal equilibrium state reached by a community in a particular environment.

Coenosteum - thin horizontal skeletal plates between corallites.

Colonial corals - corals composed of many individuals. There is no clear distinction between single individuals with many mouths and colonies of individuals with single mouths.

Commensal - a species which lives in association with another, but without harming it. Hidden.

Community - a group of organisms of different species that co-occur in the same habitat or area and interact through trophic and spatial relationships. Communities are typically characterised by reference to one or more dominant species.

Corallite - the skeleton of an individual coral polyp.

Corallum - the skeleton of a coral colony.

Dinoflagellate - are a large group of flagellate protists that constitute the phylum Dinoflagellata. Most are marine plankton. Their populations are distributed depending on temperature, salinity, or depth.

Dissepiments - blistery horizontal plates of calcium carbonate adjoining corallites.

Ecological niche - all the conditions relating to habitat, feeding regime and habits specific to a given species.

Ecomorphs - morphological variants of species that may have an environmental and/or genetic origin.

Endo-hormonos - the body has two message bearing systems, the electric rapidly reacting nervous system and a slower chemical system using the hormones.

Enteron - noun the alimentary canal (especially of an embryo or a coelenterate).

Epibiota - animals and plants living attached to or resting upon a substratum, or on another living organism.

Epibiotic - living as epibiota.

Gastropods - class of molluscs crawling around on a large ventral foot, often having a dorsal spiral shell.

Habitat – a vague word indicating the particular type of environment occupied by an organism.

Hermatypic - literally 'reef building' but commonly used as a descriptor for marine invertebrates that have photosynthetic plants living symbiotically within their tissues.

Hydrozoans - class of cnidarians including, among others, the fire coral *Millepora platyphylla*.

Meandroid - massive corals that have corallite mouths aligned in valleys such that there are no individual polyps.

Molluscs - phylum of invertebrate animals with soft bodies and generally a shell.

Photosynthetic - related to chlorophyll-linked assimilation.

Phylogeny - the evolutionary history of a group or lineage.

Phylogenesis - the evolutionary history of a taxon.

Phylogenetics - the description of evolutionary relationships using cladistic methods.

Planulae - larvae of coral.

Plocoid coral - massive corals that have corallites with separate walls.

Polychaete - segmented worm with numerous lateral bristles, belonging to the phylum Annelida.

Polyp - an individual coral including soft tissues and skeleton.

Population - a group of conspecific organisms that exhibit reproductive continuity. It is generally presumed that ecological and reproductive interactions are more frequent among members within a population than with members of other populations.

Reef flat - the flat intertidal part of reefs that are exposed to wave action.

Reef slope - the sloping part of reefs below the reef flat.

Scleractinian (corals) - corals living in symbiosis with microscopic algae, the zooxanthelae. They produce calcium carbonate in quantities sufficient to build coral reefs. Most 'hard' corals are Scleractinia.

Sessile (fauna) - attached fauna in contrast to mobile fauna (unattached).

Septa - radial skeletal elements projecting inwards from the corallite wall.

Spur-and-groove zone - morphological feature of the upper part of the outer slope, made up of ridges (spurs) aligned more or less perpendicular to the outer slope and separated by grooves.

Zoanthid - animal belonging to the phylum Cnidaria, with anemone-like appearance and no skeleton, either solitary or colonial.

Zooxanthelae - unicellular dinoflagellate algae living in the tissues of certain animals (corals and giant clams), to which they supply nutritional substances directly useable by their host.

Zooxanthellate corals - Corals that have photosynthetic endosymbiotic algae.

About the Author

Yuri Latypov

Doctor Biological Sciences, Academic New York Academy of Sciences and Russian Ecological Academy are studying the fossil and modern corals. Yuri Latypov first observed live corals in Australia about 30 years ago and was struck by the beautiful growth and diversity of these remarkable animals. Since then, he has had the opportunity to look at and study at corals and coral reefs throughout the world from Australia's Great Barrier Reef up to Seychelles and Mauritius. He spent more than 3,000 hours underwater. He has conducted fundamental and applied research on coral in many places and has published the suits of these studies in different scientific journals, 15 books (3-field guide of corals) and conference proceedings. Dr. Latypov gives special attention to the study of corals and reefs in Vietnam. Last year he investigated opportunities of restoration of reef communities and carried out experiments on artificial cultivation and rehabilitation corals on the Vietnamese reefs.

INDEX

A

B

C

D

E

F

N

O

P

R

S

T

U

V

W

Z